New Economic Windows

Series Editor
MASSIMO SALZANO

Arnab Chatterjee • Bikas K Chakrabarti (Eds.)

Econophysics of Stock and other Markets

Proceedings of the Econophys-Kolkata II

Springer

ARNAB CHATTERJEE
BIKAS K CHAKRABARTI
Theoretical Condensed Matter Physics Division and
Centre for Applied Mathematics and Computational Science
Saha Institute of Nuclear Physics, Kolkata, India

The publication of this book has been made possible by a partial support from a fund provided by the Centre for Applied Mathematics and Computational Science, Saha Institute of Nuclear Physics, Kolkata, India

Library of Congress Control Number: 2006928743

ISBN-10 88-470-0501-9 Springer Milan Berlin Heidelberg New York

ISBN-13 978-88-470-0501-3

Springer is a part of Springer Science+Business Media

springer.com

Printed in Italy

Cover design: Simona Colombo, Milano
Typeset by the authors using a Springer Macro package
Printing and binding: Signum srl, Bollate (MI)

Printed on acid-free paper

Preface

Successful or not, we all (have to?) go to various markets and participate in their activities. Yet, so little is understood about their functionings. Efforts to model various markets are now substantial. Econophysicists have also come up recently with several innovative models and their analyses.

This book is a proceedings of the International Workshop on "*Econophysics of Stock Markets and Minority Games*", held in Kolkata during February 14-17, 2006, under the auspices of the *Centre for Applied Mathematics and Computational Science*, Saha Institute of Nuclear Physics, Kolkata. This is the second event in the Econophys-Kolkata series of meetings; the Econophys-Kolkata I was held in March 2005 (Proceedings: *Econophysics of Wealth Distributions*, published in the same New Economic Windows series by Springer, Milan in 2005). We understand from the enthusiastic response of the participants that the one-day trip to the Sunderbans (Tiger Reserve; a world heritage point) along with the lecture-sessions on the vessel had been hugely enjoyable and successful. The concluding session had again very lively discussions on the workshop topics as well as on econophysics in general, initiated by J. Barkley Rosser, Matteo Marsili, Rosario Mantegna and Robin Stinchcombe (Chair). We plan to hold the next meeting in this series, on "*Econophysics and Sociophysics: Debates on Complexity Issues in Economics and Sociology*" early next year.

We are very happy that several leading economists and physicists engaged in these recent developments in the econophysics of markets, their analysis and modelling could come and participate. Although a few of them (Fabrizio Lillo, Thomas Lux and Rosario Mantegna) could not contribute to this proceedings volume due to shortage of time (we again try to get this proceedings published within six months from the workshop), we are indeed very happy that most of the invited participants could contribute in this book. The papers on market analysis and modellings are very original and their timely appearance here will render the book extremely useful for the researchers. The two historical notes and the Comments and Discussions section will give the readers two examples of personal perspectives regarding the new developments in econophysics, and some 'touch' of the lively debates taking place in these Econophys-Kolkata series of workshops.

We are extremely grateful to Mauro Gallegati and Massimo Salzano of the editorial board of this New Economic Windows series of Springer for their encouragement and support, and to Marina Forlizzi for her efficient maintenance of publication schedule.

Kolkata,
June 2006

Arnab Chatterjee
Bikas K. Chakrabarti

Contents

Part I Markets and their Analysis

On Stock-Price Fluctuations in the Periods of Booms and Stagnations
Taisei Kaizoji 3

An Outlook on Correlations in Stock Prices
Anirban Chakraborti 13

The Power (Law) of Indian Markets: Analysing NSE and BSE Trading Statistics
Sitabhra Sinha, Raj Kumar Pan 24

A Random Matrix Approach To Volatility In An Indian Financial Market
V. Kulkarni, N. Deo 35

Why do Hurst Exponents of Traded Value Increase as the Logarithm of Company Size?
Zoltán Eisler, János Kertész 49

Statistical Distribution of Stock Returns Runs
Honggang Li, Yan Gao 59

Fluctuation Dynamics of Exchange Rates on Indian Financial Market
A. Sarkar, P. Barat 67

Noise Trading in an Emerging Market: Evidence and Analysis
Debasis Bagchi 77

How Random is the Walk: Efficiency of Indian Stock and Futures Markets
Udayan Kumar Basu 85

Part II Markets and their Models

Models of Financial Market Information Ecology
Damien Challet 101

Estimating Phenomenological Parameters in Multi-Assets Markets
Giacomo Raffaelli, Matteo Marsili 113

Agents Play Mix-game
Chengling Gou 123

Triangular Arbitrage as an Interaction in Foreign Exchange Markets
Yukihiro Aiba, Naomichi Hatano 133

Modelling Limit Order Financial Markets
Robin Stinchcombe 143

Two Fractal Overlap Time Series and Anticipation of Market Crashes
Bikas K. Chakrabarti, Arnab Chatterjee, Pratip Bhattacharyya 153

The Apparent Madness of Crowds: Irrational Collective Behavior Emerging from Interactions among Rational Agents
Sitabhra Sinha 159

Agent-Based Modelling with Wavelets and an Evolutionary Artificial Neural Network: Applications to CAC 40 Forecasting
Serge Hayward 163

Information Extraction in Scheduling Problems with Non-Identical Machines
Manipushpak Mitra 175

Modelling Financial Time Series
P. Manimaran, J.C. Parikh, P.K. Panigrahi S. Basu, C.M. Kishtawal, M.B. Porecha 183

Random Matrix Approach to Fluctuations and Scaling in Complex Systems
M. S. Santhanam 192

The Economic Efficiency of Financial Markets
Yougui Wang 201

Regional Inequality
Abhirup Sarkar 208

Part III Historical Notes

A Brief History of Economics: An Outsider's Account
Bikas K Chakrabarti 219

The Nature and Future of Econophysics
J. Barkley Rosser, Jr. 225

Part IV Comments and Discussions

Econophys-Kolkata II Workshop Summary
J. Barkley Rosser, Jr. 237

Econophysics: Some Thoughts on Theoretical Perspectives
Matteo Marsili 240

Comments on "Worrying Trends in Econophysics": Income Distribution Models
Peter Richmond, Bikas K. Chakrabarti, Arnab Chatterjee, John Angle . . 244

List of Invited Speakers and Contributors

Yukihiro Aiba
Department of Physics
University of Tokyo, Komaba
Meguro, Tokyo 153-8505 Japan
aiba@iis.u-tokyo.ac.jp

John Angle
Inequality Process Institute
P.O. Box 429 Cabin John
Maryland 20818-0429 USA
angle@inequalityprocess.org

Debasis Bagchi
The Institute of Cost and Works
Accountants of India,
12, Sudder Street
Kolkata 700 016 India
dbagchi2002@yahoo.co.in

P. Barat
Variable Energy Cyclotron Centre
1/AF Bidhan Nagar
Kolkata 700064 India
pbarat@veccal.ernet.in

J. Barkley Rosser, Jr.
Department of Economics
James Madison University
Harrisonburg, VA, USA
rosserjb@jmu.edu

S. Basu
Space Applications Centre
Ahmedabad 380 015 India

Udayan Kumar Basu
Future Business School
Kolkata, India
udayan_basu@yahoo.co.in

Pratip Bhattacharyya
Physics Department
Gurudas College, Narkeldanga
Kolkata 700 054 India
pratip.bhattacharyya@saha.ac.in

Damien Challet
Nomura Centre for Quantitative
Finance
Mathematical Institute
Oxford University
24–29 St Giles'
Oxford OX1 3LB, UK
challet@maths.ox.ac.uk

Bikas K. Chakrabarti
Theoretical Condensed Matter
Physics Division and Centre for
Applied Mathematics and Computational Science
Saha Institute of Nuclear Physics
1/AF Bidhannagar
Kolkata 700064 India
bikask.chakrabarti@saha.ac.in

Anirban Chakraborti
Department of Physics
Banaras Hindu University
Varanasi 221005 India
achakraborti@yahoo.com

Arnab Chatterjee
Theoretical Condensed Matter Physics Division and Centre for Applied Mathematics and Computational Science
Saha Institute of Nuclear Physics
1/AF Bidhannagar
Kolkata 700064 India
arnab.chatterjee@saha.ac.in

N. Deo
Department of Physics and Astrophysics
University of Delhi
Delhi 110007 India
ndeo@physics.du.ac.in

Zoltán Eisler
Department of Theoretical Physics
Budapest University of Technology and Economics
Budafoki út 8
H-1111 Budapest Hungary
eisler@maxwell.phys.bme.hu

Yan Gao
Department of Systems Science
School of Management
Beijing Normal University
Beijing 100875 China
haoy@163.com

Chengling Gou
Physics Department
Beijing University of Aeronautics and Astronautics
37 Xueyuan Road, Heidian District
Beijing 100083 China
gouchengling@buaa.edu.cn

Naomichi Hatano
Institute of Industrial Science
University of Tokyo
Komaba, Meguro
Tokyo 153-8505 Japan
hatano@iis.u-tokyo.ac.jp

Serge Hayward
Department of Finance
Ecole Supérieure de Commerce de Dijon,
29, rue Sambin
21000 Dijon France
shayward@escdijon.com

Taisei Kaizoji
Division of Social Sciences
International Christian University
Tokyo 181-8585 Japan
kaizoji@icu.ac.jp

János Kertész
Department of Theoretical Physics
Budapest University of Technology and Economics
Budafoki út 8
H-1111 Budapest, Hungary
kertesz@phy.bme.hu

C. M. Kishtawal
Space Applications Centre
Ahmedabad 380 015 India

V. Kulkarni
Department of Physics and Astrophysics
University of Delhi
Delhi 110007 India
varsha_s_kulkarni@yahoo.com

Honggang Li
Department of Systems Science
School of Management
Beijing Normal University
Beijing 100875 China
hli@bnu.edu.cn

Fabrizio Lillo
Dipartimento di Fisica e Tecnologie Relative
viale delle Scienze - Edificio 18
I-90128 Palermo Italy
lillo@lagash.dft.unipa.it

Thomas Lux
Department of Economics
University of Kiel
Olshausenstr. 40
D-24118 Kiel Germany
lux@bwl.uni-kiel.de

P. Manimaran
Physical Research Laboratory
Ahmedabad 380 009 India

Rosario N. Mantegna
Dip. di Fisica e Tecnologie Relative
Universita' di Palermo
Viale delle Scienze
I-90128 Palermo Italy
mantegna@unipa.it

Matteo Marsili
The Abdus Salam International Centre for Theoretical Physics
Strada Costiera 14
34014 Trieste Italy
marsili@ictp.trieste.it

Manipushpak Mitra
Economic Research Unit
Indian Statistical Institute
203 B. T. Road
Kolkata-700108 India
mmitra@isical.ac.in

Raj Kumar Pan
The Institute of Mathematical Sciences
C. I. T. Campus, Taramani
Chennai 600 113 India
rajkp@imsc.res.in

P. K. Panigrahi
Physical Research Laboratory
Ahmedabad 380 009 India

J. C. Parikh
Physical Research Laboratory
Ahmedabad 380 009 India
parikh@prl.res.in

M. B. Porecha
Physical Research Laboratory
Ahmedabad 380 009 India

Giacomo Raffaelli
Dipartimento di Fisica
CNR-ISC and INFM-SMC
Universitá di Roma "La Sapienza"
p.le A. Moro 2, 00185 Roma Italy

Peter Richmond
School of Physics
University of Dublin
Trinity College Dublin 2 Ireland
richmond@tcd.ie

M. S. Santhanam
Physical Research Laboratory
Navarangpura
Ahmedabad 380 009 India
santh@prl.res.in

Abhirup Sarkar
Economic Research Unit
Indian Statistical Institute
203 B. T. Road
Kolkata 700108 India
abhirup@isical.ac.in

Apu Sarkar
Variable Energy Cyclotron Centre
1/AF Bidhan Nagar
Kolkata 700064 India
apu@veccal.ernet.in

Sitabhra Sinha
The Institute of Mathematical Sciences
C. I. T. Campus, Taramani
Chennai 600 113 India
sitabhra@imsc.res.in

Robin Stinchcombe
Rudolf Peierls Centre for Theoretical Physics
Oxford University
1 Keble Road
OXFORD, OX1 3NP, UK
stinch@thphys.ox.ac.uk

Yougui Wang
Department of Systems Science
School of Management
Beijing Normal University
Beijing 100875
People's Republic of China
ygwang@bnu.edu.cn

Part I

Markets and their Analysis

On Stock-Price Fluctuations in the Periods of Booms and Stagnations

Taisei Kaizoji

Division of Social Sciences, International Christian University, Tokyo 181-8585, Japan. kaizoji@icu.ac.jp

1 Introduction

The statistical properties of the fluctuations of financial prices have been widely researched since Mandelbrot [1] and Fama [2] presented an evidence that return distributions can be well described by a symmetric Levy stable law with tail index close to 1.7. Many empirical studies have shown that the tails of the distributions of returns and volatility follow approximately a power law with estimates of the tail index falling in the range 2 to 4 for large value of returns and volatility. (See, for examples, de Vries [3]; Pagan [4]; Longin [5], Lux [6]; Guillaume et al. [7]; Muller et al. [8]; Gopikrishnan et al. [9], Gopikrishnan et al. [10], Plerou et al. [11], Liu et al. [12]). However, there is also evidence against power-law tails. For instance, Barndorff-Nielsen [13], and Eberlein et al. [14] have respectively fitted the distributions of returns using normal inverse Gaussian, and hyperbolic distribution. Laherrere and Sornette [15] have suggested to describe the distributions of returns by the stretched-exponential distribution. Dragulescu and Yakovenko [16] have shown that the distributions of returns have been approximated by exponential distributions. More recently, Malevergne, Pisarenko and Sornette [17] have suggested that the tails ultimately decay slower than any stretched exponential distribution but probably faster than power laws with reasonable exponents as a result from various statistical tests of returns.

Thus opinions vary among scientists as to the shape of the tail of the distribution of returns (and volatility). While there is fairly general agreement that the distribution of returns and volatility has fat tails for large values of returns and volatility, there is still room for a considerable measure of disagreement about universality of the power-law distributions. At the moment we can only say with fair certainty that (i) the power-law tail of the distribution of returns and volatility is not an universal law and (ii) the tails of the distribution of returns and volatility are heavier than a Gaussian, and are between exponential and power-law. There is one other thing that is important for understanding of price movements in financial markets. It is a fact that the financial market has repeated booms (or bull market) and stagnations (or bear market). To ignore this fact is to miss the reason why price fluctuations are caused. However, in most empirical studies, which have been made on statistical properties of returns and volatility in financial markets, little attention has been given to the relationship between market situations and price fluctuations. Our previous work [18] investigates this subject using the historical data of the Nikkei 225 index. We find that the volatility in the inflationary period is approximated by an *power-law* distribution while the volatility distribution in the deflationary period is described by an *exponential* distribution. The purpose of this paper is to examine further the statistical properties of volatility distribution from this viewpoint. We use the daily data of the four stock price indices of the three major stock markets in the world: the Nikkei 225 index, the DJIA. SP500, and FT100, and compare the shape of the volatility distribution for each of the stock price indices in the periods of booms with that in the period of stagnations. We find that (i) the tails of the distribution of the absolute log-returns are approximated by a power-law function with the exponent close to 3 in the periods of booms while the distribution is described by an exponential function with the scale parameter close to unity in the periods of stagnations. These indicate that so far as the stock price indices we used are concerned, the same observation on the volatility distribution holds in all cases.

The rest of the paper is organized as follows: the next section analyzes the stock price indices and shows the empirical findings. Section 3 gives concluding remarks.

2 Empirical analysis

2.1 Stock price indices

We investigate quantitatively the four stock price indices of the three major stock markets in the world[1], that is, (a) the Nikkei 225 index (Nikkei 225), which is the price-weighted average of the stock prices for 225 large companies listed in the Tokyo Stock Exchange, (b) the Dow Jones Industrial Average (DJIA), which is the price-weighted average of 30 significant stocks traded on the New York Stock Exchange and Nasdaq, (c) Standard and Proor's 500 index (SP 500), which is a market-value weighted index of 500 stocks chosen for market size, liquidity, and industry group representation, and (d) FT 100, which is similar to SP 500, and a market-value weighted index of shares of the top 100 UK companies ranked by market capitalization. Figure 1(a)-(d) show the daily series of the four stock price indices: (a) the Nikkei 225 from January 4, 1975 to August 18, 2004, (b) DJIA from January 2, 1946 to August

[1] The prices of the indices are close prices which are adjusted for dividends and splits.

18, 2004, (c) SP 500 from November 22, 1982 to August 18, 2004, and (d) FT 100 from April 2, 1984 to August 18, 2004.

After booms of a long period of time, the Nikkei 225 reached a high of almost 40,000 yen on the last trading day of the decade of the 1980s, and then from the beginning trading day of 1990 to mid-August 1992, the index had declined to 14,309, a drop of about 63 percent. A prolonged stagnation of the Japanese stock market started from the beginning of 1990. The time series of the DJIA and SP500 had the apparent positive trends until the beginning of 2000. Particularly these indices surged from the mid-1990s. There is no doubt that this stock market booms in history were propelled by the phenomenal growth of the Internet which has added a whole new stratum of industry to the American economy. However, the stock market booms in the US stock markets collapsed at the beginning of 2000, and the descent of the US markets started. The DJIA peaked at 11722.98 on January 14, 2000, and dropped to 7286.27 on October 9, 2002 by 38 percent. SP500 arrived at peak for 1527.46 on March 24, 2000 and hit the bottom for 776.76 on October 10, 2002. SP500 dropped by 50 percent. Similarly FT100 reached a high of 6930.2 on December 30, 2000 and the descent started from the time. FT100 dropped to 3287 on March 12, 2003 by 53 percent.

From these observations we divide the time series of these indices in the two periods on the day of the highest value. We define the period a period until it reaches the highest value as the period of booms and the period after that as stagnations, respectively. The periods of booms and stagnations for each index of the four indices are collected into Table 1.

Table 1. The periods of booms and stagnations.

Name of Index	The Period of Booms	The Period of Stagnations
Nikkei225	4 Jan 1975-30 Dec 1989	4 Jan. 1990-18 Aug 2004
DJIA	2 Jan 1946-14 Jan 2000	18 Jan 2000-18 Aug 2004
SP500	22 Nov 1982-24 Mar 2000	27 Mar 2000-18 Aug 2004
FT100	3 Mar 1984-30 Dec 1999	4 Jan 2000-18 Aug 2004

2.2 Comparisons of the distributions of absolute log returns

In this paper we investigate the shape of distributions of absolute log returns of the stock price indices. We concentrate to compare the shape of the distribution of volatility in the period of booms with that in the period of stagnations. We use absolute log return, which is a typical measure of volatility. The absolute log returns is defined as $|R(t)| = |lnS(t) - lnS(t-1)|$, where $S(t)$ denotes the index at date t. We normalize the absolute log-return $|R(t)|$ using the standard deviation. The normalized absolute log return $V(t)$ is defined as

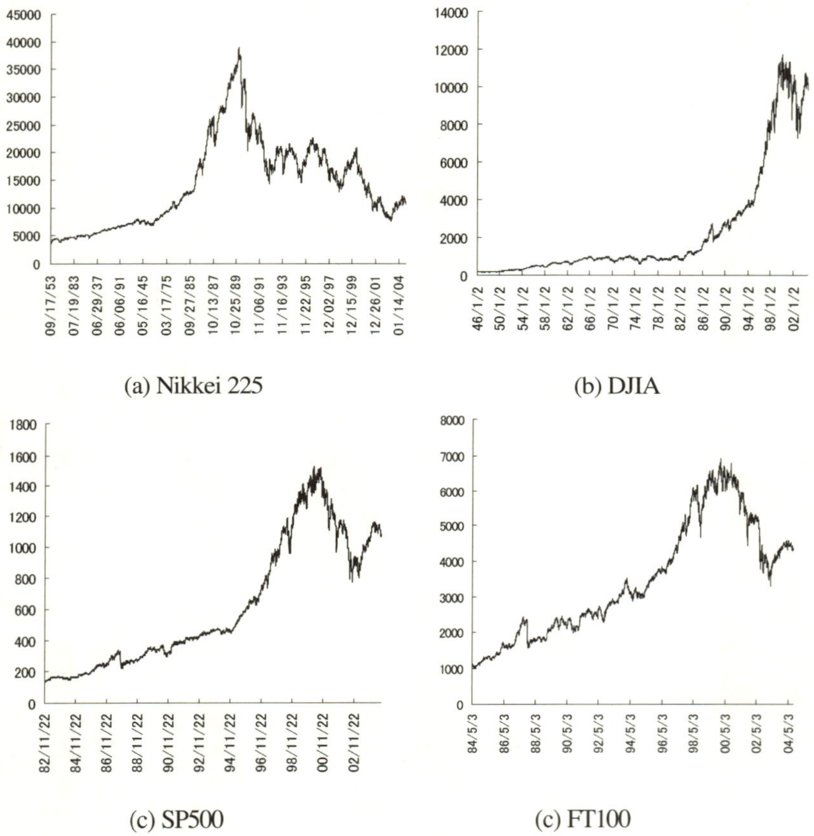

Fig. 1. The movements of the stock price indices: (a) Nikkei 225 (b) DJIA, (c) SP500, (d) FT100.

$V(t) = |R(t)|/\sigma$ where σ denotes the standard deviation of $|R(t)|$. The panels (a)-(h) of Figure 2 show the semi-log plots of the complementary cumulative distribution of the normalized absolute log-returns V for each of the four indices: Nikkei225, DJIA, SP500, and FT100 in the period of booms and that in the period of stagnations, respectively. In the all panels it follows that the tail of the volatility distribution of V is heavier in the period of booms than in the period of stagnations. The solid lines in all panels represent the fits of the exponential distribution,

$$P(V > x) \propto exp(-\frac{x}{\beta}) \tag{1}$$

where the scale parameter β is estimated from the data using a least squared method. In all cases of the period of stagnations, which are panels (b), (d), (f) and (h), the exponential distribution (1) describes very well the distributions of V over a whole range of values of V[2]. The scale parameter β is estimated from the data except for these two extreme values using a least squared method is collected in Table 2. In all cases the estimated values β are very close to unity.

Table 2. The scale parameter β of an exponential function (1) estimated from the data using the least squared method. R^2 denotes the coefficient of determinant.

Name of Index	The scale parameter β	R^2
Nikkei225	1.02	0.995
DJIA	1.09	0.995
SP500	0.99	0.997
FT100	0.99	0.999

On the other hand, the panels (a), (c), (e), and (g) of Figure 2 show the complementary cumulative distribution of V in the period of booms for each of the four indices in the semi-log plots. The solid lines in all panels represent the fits of the exponential distribution estimated from the data of only the low values of V using a least squared method. In these cases the low values of V are only approximately well described by the exponential distribution (1), but completely fails in describing the large values of V. Apparently, an exponential distribution underestimates large values in the complementary cumulative distribution of V in the period of booms.

Finally the panels (a) and (b) of Figure 3 show the complementary cumulative distributions of V for the four indices in the period of booms in a log-log

[2] Note that we exclude the tow extreme values of V: (i) the extreme value in the Nikkei 225 on September 28, 1990, and (ii) the extreme value in the DJIA on September 10, 2001. The jump of Nikkei 225 perhaps was caused by investors speculation on the 1990 Gulf War. The extreme value of DJIA was caused by terror attack in New York on September 10, 2001.

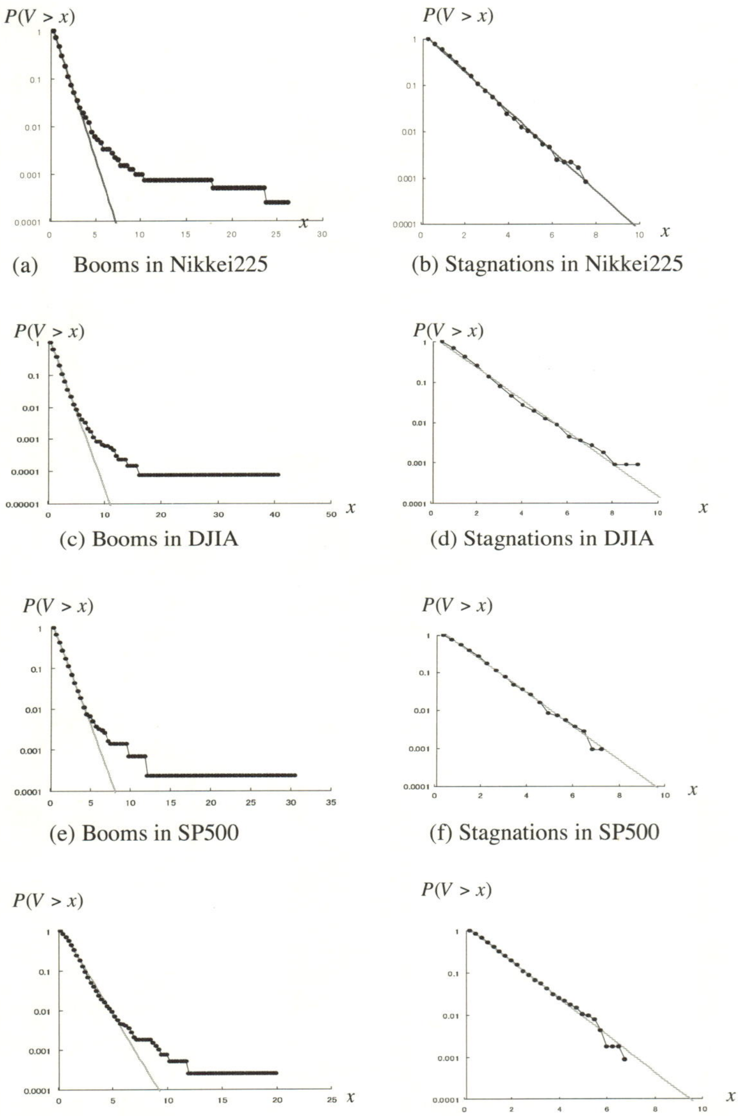

(a) Booms in Nikkei225

(b) Stagnations in Nikkei225

(c) Booms in DJIA

(d) Stagnations in DJIA

(e) Booms in SP500

(f) Stagnations in SP500

(g) Booms in FT100

(h) Stagnations in FT100

Fig. 2. The panels (a), (c), (e) and (g) indicate the complementary cumulative distribution of absolute log returns V for each of the four stock price indices in the period of booms, and the panels (b), (d), (f) and (h) indicate that in the period of stagnations. These figures are shown in a semi-log scale. The solid lines represent fits of the exponential distribution.

scale, and those in the period of stagnations in a semi-log scale. The two figures confirm that the shape of the fourth volatility distributions in the periods of booms and of stagnations is almost the same, respectively. Furthermore, the complementary cumulative distribution of V in the period of booms for each of the four indices are approximately described by the power-law distribution in the large values of V,

$$P(V > x) \propto x^{-\alpha} \tag{2}$$

The power-law exponent α is estimated from the data of the large values of V using the least squared method. The best fits succeed in describing approximately large values of V. Table 3 collect the power-law exponent α, which is estimated. The estimated value α are in the range from 2.83 to 3.69.

Table 3. The power-law exponent α of a power-law function (2) estimated from the data using the least squared method. R^2 denotes the coefficient of determinant.

Name of Index	The power-law parameter α	R^2
Nikkei225	2.83	0.992
DJIA	3.69	0.995
SP500	3.26	0.986
FT100	3.16	0.986

3 Concluding remarks

In this paper we focus on comparisons of shape of the distributions of absolute log returns in the period of booms with those in the period of stagnations for the four major stock price indices. We find that the complementary cumulative distribution in the period of booms is very well described by exponential distribution with the scale parameter close to unity while the complementary cumulative distribution of the absolute log returns is approximated by power-law distribution with the exponent in the range of 2.8 to 3.8. The latter is complete agreement with numerous evidences to show that the tail of the distribution of returns and volatility for large values of volatility follow approximately a power law with the estimates of the exponent falling in the range 2 to 4. We are now able to see that the statistical properties of volatility for stock price index are changed according to situations of the stock markets. Our findings make it clear that we must look more carefully into the relationship between regimes of markets and volatility in order to fully understand price fluctuations in financial markets. The question, which we must consider next, is the reasons why and how the differences are created. That traders herd behavior may help account for it would be accepted by most people. Recently we have proposed a stochastic model [19] that may offer the key

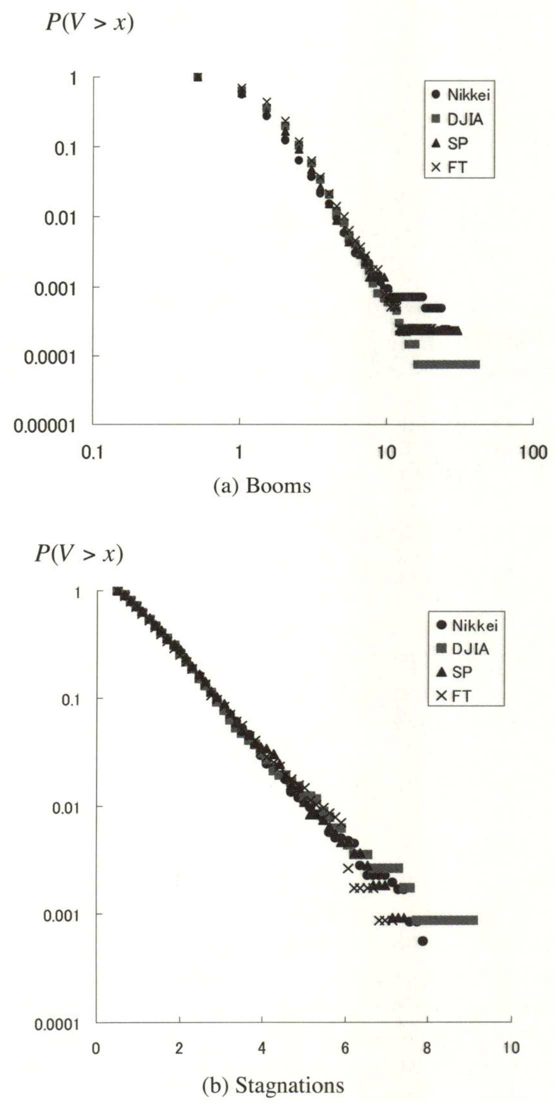

Fig. 3. The panels (a) and (b) show the complementary cumulative distributions of V for the four indices in the period of booms in a log-log scale, and those in the period of stagnations in a semi-log scale.

to an understanding of the empirical findings we present here. The results of the numerical simulation of the model suggest the following: in the period of booms, the noise traders' herd behavior strongly influences to the stock market and generate power-law tails of the volatility distribution while in the period of stagnations a large number of noise traders leave a stock market and interplay with the noise traders become weak, so that exponential tails of the volatility distribution is observed. However it remains an unsettled question what causes switching from boom to stagnation[3]. Our findings may provide a starting point to make a new tool of risk management of index fund in financial markets, but to apply the rule which we show here to risk management, we need to establish the framework of analysis and refine the statistical methods. We began with a simple observation on the stock price indices, and divided the price series into the two periods: booms and stagnations. However, there is room for further investigation on how to split the price series into periods according to the situations of markets.

Acknowledgements

My special thanks are due to Prof. Thomas Lux and Prof. Enrico Scalas for valuable comments, and the organizeer of Econophysics-Kolkata II, Prof. Bikas K Chakrabarti.

An earlier version of this paper was presented at the 8th Annual Workshop on Economics with Heterogeneous Interacting Agents (WEHIA2003) held at Institute in World Economy, Kiel, Germany, May 29-31, 2003. Financial support by the Japan Society for the Promotion of Science under the Grant-in-Aid is gratefully acknowledged.

References

1. Mandelbrot B (1963) Journal of Business 36:392-417
2. Fama EF (1965) Journal of Business 38:34-105
3. de Vries CG (1994) in The Handbook of International Macroeconomics, F. van der Ploeg (ed.) pp.348-389 Blackwell
4. Pagan A (1996) Journal of Empirical Finance 3:15-102
5. Longin FM (1996) Journal of Business 96:383-408

[3] Yang, et.al. [20] also studies the log-return the dynamics of the log-return distribution of the Korean Composition Stock Price Index (KOSPI) from 1992 to 2004. As a result of the empirical study using intraday data of the index, they found that while the index during the late 1990s showed a power-law distribution, the distribution in the early 2000s was exponential. To explain this change in distribution shape, they propose a spin like model of financial markets. They show that changing the shape of the return distribution was caused by changing the transmission speeds of the information that flowed into the market.

6. Lux T (1996) Applied Financial Economics 6:463-475
7. Guillaume DM, Dacorogna MM, Dave RR, Muller UA, Olsen RB, Pictet OV (1997) Finance and Stochastics 1:95-130
8. Muller UA, Dacarogna MM, Picktet OV (1998) in R.J.Adler, R.E.Feldman, M.S.Taqqu, Birkhauser, Boston (eds.), A Practical Guide to Heavy Tails, pp.55-78
9. Gopikrishnan P, Meyer M, Amaral LAN, Stanley HE (1998) European Physical Journal B 3:139-140
10. Gopikrishnan P, Plerou V, Amaral LAN, Meyer M, Stanley HE (1999) Phys. Rev. E 60:5305-5316
11. Plerou V, Gopikrishnan P, Amaral LAN, Meyer M, Stanley HE (2000) Phys. Rev. E 60:6519-6529
12. Liu Y, Gopikrishnan P, Cizeau P, Meyer M, Peng C-K, Stanley HE (1990) Physical Review E 60(2):1390-1400
13. Barndorff-Nielsen OE (1997) Scandinavian Journal of Statistics 24:1-13
14. Eberlein E, Keller U, Prause K (1998) Journal of Business 71:371-405
15. Laherrere J, Sornette D (1999) European Physical Journal B 2:525-539
16. Dragulescu AA, Yakovenko VM (2002) Quantitative Finance 2(6):443-453
17. Malevergne Y, Pisarenko VF, Sornette D (2005) Quantitative Finance 5(4):379-401
18. Kaizoji T (2004), Physica A343:662-668.
19. Kaizoji T (2005) in T. Lux, S. Reitz, and E. Samanidou (eds.): Nonlinear Dynamics and Heterogeneous Intercting Agents: Lecture Notes in Economics and Mathematical Systems 550, Springer-Verlag, Berlin- Heidelberg, pp.237–248
20. Yang J-S, Chae S, Jung W-S, Moon H-T (2006) Physica A 363:377-382

An Outlook on Correlations in Stock Prices

Anirban Chakraborti

Department of Physics, Banaras Hindu University, Varanasi-221005, India.
achakraborti@yahoo.com

Summary. We present an outlook of the studies on correlations in the price time-series of stocks, discussing the construction and applications of "asset tree". The topic discussed here should illustrate how the complex economic system (financial market) enrichens the list of existing dynamical systems that physicists have been studying for long.

> *"If stock market experts were so expert, they would be buying stock, not selling advice."*
> – Norman Augustine, US aircraft businessman (1935 -)

1 Introduction

The word "correlation" is defined as "a relation existing between phenomena or things or between mathematical or statistical variables which tend to vary, be associated, or occur together in a way not expected on the basis of chance alone" (see http://www.m-w.com/dictionary/correlations). As soon as we talk about "chance", the words "probability","random", etc come to our mind. So, when we talk about correlations in stock prices, what we are really interested in are the nature of the time series of stock prices, the relation of stock prices with other variables like stock transaction volumes, the statistical distributions and laws which govern the price time series, in particular whether the time series is random or not. The first formal efforts in this direction were those of Louis Bachelier, more than a century ago [1]. Eversince, financial time series analysis is of prevalent interest to theoreticians for making inferences and predictions though it is primarily an empirical discipline. The uncertainty in the financial time series and its theory makes it specially interesting to statistical physicists, besides financial economists [2,3]. One of the most debatable issues in financial economics is whether the market is "efficient" or not. The "efficient" asset market is one in which the information contained in past prices is instantly, fully and continually reflected in the asset's current price. As a consequence, the more efficient the market is, the more random is the sequence of price changes generated by the market. Hence, the most efficient market is one in which the price changes are completely random and unpredictable. This leads to another relevant or pertinent question of financial econometrics: whether asset prices are predictable. Two simplest models of probability theory and financial econometrics that deal with predicting future price changes, the random walk theory and Martingale theory, assume that the future price changes are functions of only the past price changes. Now, in Economics the "logarithmic returns" is calculated using the formula

$$r(t) = \ln P(t) - \ln P(t-1), \tag{1}$$

where $P(t)$ is the price (index) at time step t. A main characteristic of the random walk and Martingale models is that the returns are *uncorrelated*.

In the past, several hypotheses have been proposed to model financial time series and studies have been conducted to explain their most characteristic features. The study of long-time correlations in the financial time series is a very interesting and widely studied problem, especially since they give a deep insight about the underlying processes that generate the time series [4]. The complex nature of financial time series (see Fig. 1) has especially forced the physicists to add this system to their existing list of dynamical systems that they study. Here, we will not try to review all the studies, but instead give a brief outlook of the studies done by the author and his collaborators, and the motivated readers are kindly asked to refer the original papers for further details.

2 Analysing correlations in stock price time series

2.1 Financial correlation matrix and constructing asset trees

In our studies, we used two different sets of financial data for different purposes. The first set from the Standard & Poor's 500 index (S&P500) of the New York Stock Exchange (NYSE) from July 2, 1962 to December 31, 1997 containing 8939 daily closing values, which we have already plotted in Fig. 1(d). In the second set, we study the split-adjusted daily closure prices for a total of $N = 477$ stocks traded at the New York Stock Exchange (NYSE) over the period of 20 years, from 02-Jan-1980 to 31-Dec-1999. This amounts a total of 5056 price quotes per stock, indexed by time variable $\tau = 1, 2, \ldots, 5056$. For analysis and smoothing purposes, the data is divided time-wise into M *windows* $t = 1, 2, \ldots, M$ of width T, where T corresponds to the number of daily returns included in the window. Several consecutive windows overlap with each other, the extent of which is dictated by the window step length parameter δT, which describes the displacement of the window and is also measured in trading days. The choice of window width is a trade-off between too noisy and too smoothed data for small and large window widths, respectively. The results presented in this paper were calculated from monthly stepped four-year windows, i.e. $\delta T = 250/12 \approx 21$ days and $T = 1000$ days. We have explored a large scale of different values for both parameters, and the cited values were found optimal [5]. With these choices, the overall number of windows is $M = 195$.

In order to investigate correlations between stocks we first denote the closure price of stock i at time τ by $P_i(\tau)$ (Note that τ refers to a date, not a time window). We focus our attention to the logarithmic return of stock i, given by $r_i(\tau) = \ln P_i(\tau) - \ln P_i(\tau - 1)$ which for a sequence of consecutive trading days, i.e. those encompassing the given window t, form the return vector $\boldsymbol{r}_i^t$. In order to characterize the synchronous time evolution of assets,

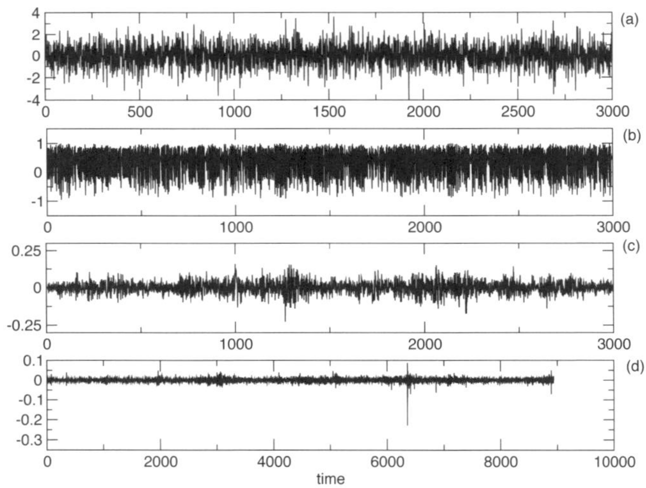

Fig. 1. Comparison of several time series which are of interest to physicists and economists: (a) Random time series (3000 time steps) using random numbers from a Normal distribution with zero mean and unit standard deviation. (b) Multivariate spatio-temporal time series (3000 time steps) drawn from the class of diffusively coupled map lattices in one-dimension with sites $i = 1, 2...n'$ of the form: $y_{t+1}^i = (1-\epsilon)f(y_t^i) + \frac{\epsilon}{2}(f(y_t^{i+1}) + f(y_t^{i-1}))$, where $f(y) = 1 - ay^2$ is the logistic map whose dynamics is controlled by the parameter a, and the parameter ϵ is a measure of coupling between nearest-neighbor lattice sites. We use parameters $a = 1.97$, $\epsilon = 0.4$ for the dynamics to be in the regime of spatio-temporal chaos. We choose $n = 500$ and iterate, starting from random initial conditions, for $p = 5 \times 10^7$ time steps, after discarding 10^5 transient iterates. Also, we choose periodic boundary conditions, $x(n+1) = x(1)$. (c) Multiplicative stochastic process GARCH(1,1) for a random variable x_t with zero mean and variance σ_t^2, characterized by a Gaussian conditional probability distribution function $f_t(x)$: $\sigma_t^2 = \alpha_0 + \alpha_1 x_{t-1}^2 + \beta_1 \sigma_{t-1}^2$, using parameters $\alpha_0 = 0.00023$, $\alpha_1 = 0.09$ and $\beta_1 = 0.01$ (3000 time steps). (d) Empirical Return time series of the S&P500 stock index (8938 time steps).

we use the equal time correlation coefficients between assets i and j defined as

$$\rho_{ij}^t = \frac{\langle \boldsymbol{r}_i^t \boldsymbol{r}_j^t \rangle - \langle \boldsymbol{r}_i^t \rangle \langle \boldsymbol{r}_j^t \rangle}{\sqrt{[\langle \boldsymbol{r}_i^{t2} \rangle - \langle \boldsymbol{r}_i^t \rangle^2][\langle \boldsymbol{r}_j^{t2} \rangle - \langle \boldsymbol{r}_j^t \rangle^2]}}, \tag{2}$$

where $\langle ... \rangle$ indicates a time average over the consecutive trading days included in the return vectors. These correlation coefficients fulfill the condition $-1 \leq \rho_{ij} \leq 1$. If $\rho_{ij} = 1$, the stock price changes are completely correlated; if $\rho_{ij} = 0$, the stock price changes are uncorrelated and if $\rho_{ij} = -1$, then the stock price changes are completely anti-correlated [6]. These correlation coefficients form an $N \times N$ correlation matrix $\mathbf{C}^t$, which serves as the basis for trees discussed in this paper.

We construct an asset tree according to the methodology by Mantegna [7]. For the purpose of constructing asset trees, we define a distance between a pair of stocks. This distance is associated with the edge connecting the stocks and it is expected to reflect the level at which the stocks are correlated. We use a simple non-linear transformation $d_{ij}^t = \sqrt{2(1-\rho_{ij}^t)}$ to obtain distances with the property $2 \geq d_{ij} \geq 0$, forming an $N \times N$ symmetric distance matrix $\mathbf{D}^t$. So, if $d_{ij} = 0$, the stock price changes are completely correlated; if $d_{ij} = 2$, the stock price changes are completely anti-uncorrelated. The trees for different time windows are not independent of each other, but form a series through time. Consequently, this multitude of trees is interpreted as a sequence of evolutionary steps of a single *dynamic asset tree*. We also require an additional hypothesis about the topology of the metric space, the ultrametricity hypothesis. In practice, it leads to determining the minimum spanning tree (MST) of the distances, denoted $\mathbf{T}^t$. The spanning tree is a simply connected acyclic (no cycles) graph that connects all N nodes (stocks) with $N-1$ edges such that the sum of all edge weights, $\sum_{d_{ij}^t \in \mathbf{T}^t} d_{ij}^t$, is minimum. We refer to the minimum spanning tree at time t by the notation $\mathbf{T}^t = (V, E^t)$, where V is a set of vertices and E^t is a corresponding set of unordered pairs of vertices, or edges. Since the spanning tree criterion requires all N nodes to be always present, the set of vertices V is time independent, which is why the time superscript has been dropped from notation. The set of edges E^t, however, does depend on time, as it is expected that edge lengths in the matrix $\mathbf{D}^t$ evolve over time, and thus different edges get selected in the tree at different times.

2.2 Market characterization

We plot the distribution of (i) distance elements d_{ij}^t contained in the distance matrix $\mathbf{D}^t$ (Fig. 2), (ii) distance elements d_{ij} contained in the asset (minimum spanning) tree $\mathbf{T}^t$ (Fig. 3). In both plots, but most prominently in Fig. 2, there appears to be a discontinuity in the distribution between roughly 1986 and 1990. The part that has been cut out, pushed to the left and made flatter, is a manifestation of Black Monday (October 19, 1987), and its length along the time axis is related to the choice of window width T [6,8]. Also, note that in

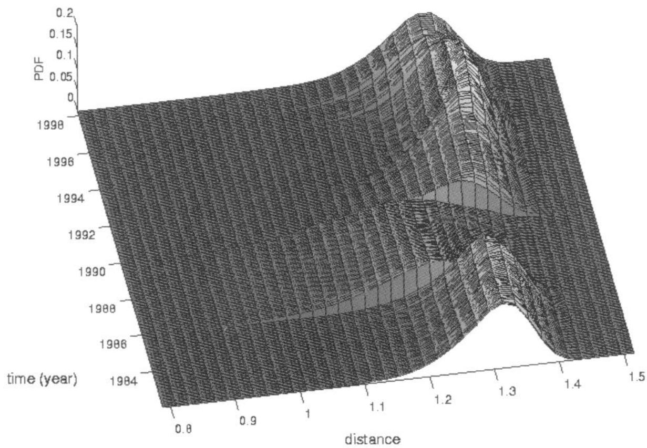

Fig. 2. Distribution of all $N(N-1)/2$ distance elements d_{ij} contained in the distance matrix $\mathbf{D}^t$ as a function of time.

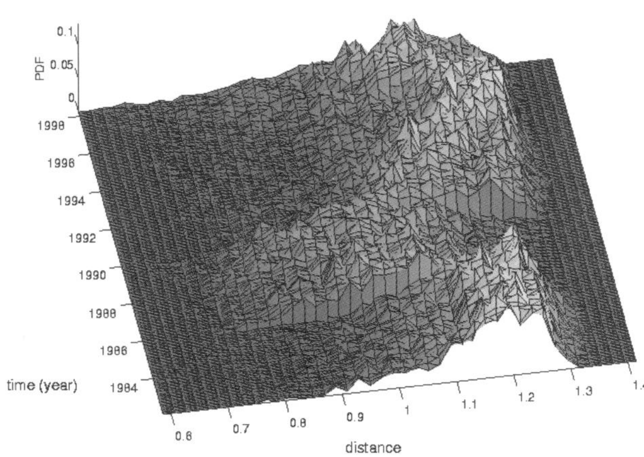

Fig. 3. Distribution of the $(N-1)$ distance elements d_{ij} contained in the asset (minimum spanning) tree $\mathbf{T}^t$ as a function of time.

the distribution of tree edges in Fig. 3 most edges included in the tree seem to come from the area to the right of the value 1.1 in Fig. 2, and the largest distance element is $d_{max} = 1.3549$.

Tree occupation and central vertex

We focus on characterizing the spread of nodes on the tree, by introducing the quantity of *mean occupation layer*

$$l(t, v_c) = \frac{1}{N} \sum_{i=1}^{N} \text{lev}(v_i^t), \tag{3}$$

where $\text{lev}(v_i)$ denotes the level of vertex v_i. The levels, not to be confused with the distances d_{ij} between nodes, are measured in natural numbers in relation to the *central vertex* v_c, whose level is taken to be zero. Here the mean occupation layer indicates the layer on which the mass of the tree, on average, is conceived to be located. The central vertex is considered to be the parent of all other nodes in the tree, and is also known as the root of the tree. It is used as the *reference* point in the tree, against which the locations of all other nodes are relative. Thus all other nodes in the tree are children of the central vertex. Although there is an *arbitrariness* in the choice of the central vertex, we propose that it is central, in the sense that any change in its price strongly affects the course of events in the market on the whole. We have proposed three alternative definitions for the central vertex in our studies, all yielding similar and, in most cases, identical outcomes. The idea here is to find the node that is most strongly connected to its nearest neighbors. For example, according to one definition, the central node is the one with the highest *vertex degree*, i.e. the number of edges which are incident with (neighbor of) the vertex. Also, one may have either (i) static (fixed at all times) or (ii) dynamic (updated at each time step) central vertex, but again the results do not seem to vary significantly. We can then study the variation of the topological properties and nature of the trees, with time. This type of visualization tool can sometimes provide deeper insight of the dynamical system.

Economic taxonomy

Mantegna's idea of linking stocks in an ultrametric space was motivated *a posteriori* by the property of such a space to provide a meaningful economic taxonomy. In [7], Mantegna examined the meaningfulness of the taxonomy by comparing the grouping of stocks in the tree with a third party reference grouping of stocks by their industry etc. classifications. In this case, the reference was provided by Forbes [9], which uses its own classification system, assigning each stock with a sector (higher level) and industry (lower level) category. In order to visualize the grouping of stocks, we constructed a sample asset tree for a smaller dataset (shown in Fig. 4) [10], which con-

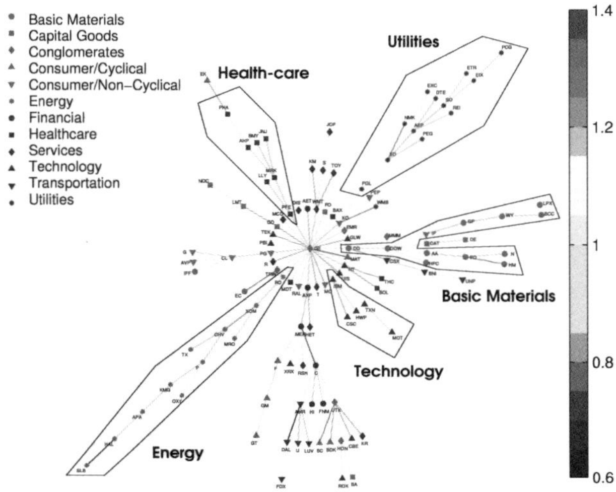

Fig. 4. Snapshot of a dynamic asset tree connecting the examined 116 stocks of the S&P 500 index. The tree was produced using four-year window width and it is centered on January 1, 1998. Business sectors are indicated according to Forbes, *http://www.forbes.com*. In this tree, General Electric (GE) was used as the central vertex and eight layers can be identified.

sists of 116 S&P 500 stocks, extending from the beginning of 1982 to the end of 2000, resulting in a total of 4787 price quotes per stock [11]. The window width was set at $T = 1000$, and the shown sample tree is located time-wise at $t = t^*$, corresponding to 1.1.1998. The stocks in this dataset fall into 12 *sectors*, which are Basic Materials, Capital Goods, Conglomerates, Consumer/Cyclical, Consumer/Non-Cyclical, Energy, Financial, Healthcare, Services, Technology, Transportation and Utilities. The sectors are indicated in the tree (see Fig. 4) with different markers, while the industry classifications are omitted for reasons of clarity. We use the term sector exclusively to refer to the given third party classification system of stocks. The term *branch* refers to a subset of the tree, to all the nodes that share the specified common parent. In addition to the parent, we need to have a reference point to indicate the generational direction (i.e. who is who's parent) in order for a branch to be well defined. Without this reference there is absolutely no way to determine where one branch ends and the other begins. In our case, the reference is the central node. There are some branches in the tree, in which most of the stocks belong to just one sector, indicating that the branch is fairly homogeneous with respect to business sectors. This finding is in accordance with those of Mantegna [7], although there are branches that are fairly heterogeneous, such as the one extending directly downwards from the central vertex (see Fig. 4).

2.3 Portfolio analysis

Next, we apply the above discussed concepts and measures to the portfolio optimization problem, a basic problem of financial analysis. This is done in the hope that the asset tree could serve as another type of quantitative approach to and/or visualization aid of the highly inter-connected market, thus acting as a tool supporting the decision making process. We consider a *general Markowitz portfolio* $\mathbf{P}(t)$ with the asset weights $w_1, w_2, \ldots, w_N$. In the classic Markowitz portfolio optimization scheme, financial assets are characterized by their average risk and return, where the risk associated with an asset is measured by the standard deviation of returns. The Markowitz optimization is usually carried out by using historical data. The aim is to optimize the asset weights so that the overall portfolio risk is minimized for a given portfolio return $r_{\mathbf{P}}$. In the dynamic asset tree framework, however, the task is to determine how the assets are located with respect to the central vertex.

Let r_m and r_M denote the returns of the minimum and maximum return portfolios, respectively. The expected portfolio return varies between these two extremes, and can be expressed as $r_{\mathbf{P},\theta} = (1-\theta)r_m + \theta r_M$, where θ is a fraction between 0 and 1. Hence, when $\theta = 0$, we have the minimum risk portfolio, and when $\theta = 1$, we have the maximum return (maximum risk) portfolio. The higher the value of θ, the higher the expected portfolio return $r_{\mathbf{P},\theta}$ and, consequently, the higher the risk the investor is willing to absorb. We define a single measure, the *weighted portfolio layer* as

$$l_{\mathbf{P}}(t,\theta) = \sum_{i \in \mathbf{P}(t,\theta)} w_i \operatorname{lev}(v_i^t), \tag{4}$$

where $\sum_{i=1}^{N} w_i = 1$ and further, as a starting point, the constraint $w_i \geq 0$ for all i, which is equivalent to assuming that there is no short-selling. The purpose of this constraint is to prevent negative values for $l_{\mathbf{P}}(t)$, which would not have a meaningful interpretation in our framework of trees with central vertex. This restriction will shortly be discuss further.

Fig. 5 shows the behavior of the mean occupation layer $l(t)$ and the weighted minimum risk portfolio layer $l_{\mathbf{P}}(t,\theta=0)$. We find that the portfolio layer is higher than the mean layer at all times. The difference between the layers depends on the window width, here set at $T = 1000$, and the type of central vertex used. The upper plot in Fig. 5 is produced using the static central vertex (GE), and the difference in layers is found to be 1.47. The lower one is produced by using a dynamic central vertex, selected with the vertex degree criterion, in which case the difference of 1.39 is found. Here, we had assumed the no short-selling condition. However, it turns out that, in practice, the weighted portfolio layer never assumes negative values and the short-selling condition, in fact, is not necessary. Only minor differences are observed in the results between banning and allowing short-selling. Further, the difference in layers is also slightly larger for static than dynamic central vertex, although not by a significant amount.

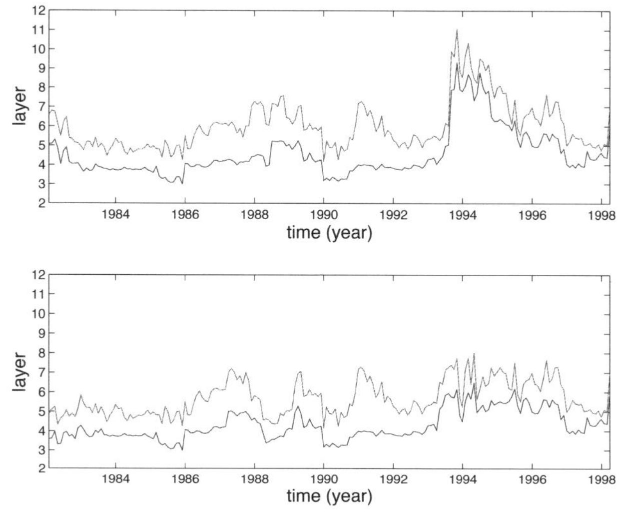

Fig. 5. Plot of the weighted minimum risk portfolio layer $l_{\mathbf{P}}(t, \theta = 0)$ with no short-selling and mean occupation layer $l(t, v_c)$ against time. Top: static central vertex, bottom: dynamic central vertex according to the vertex degree criterion.

As the stocks of the minimum risk portfolio are found on the outskirts of the tree, we expect larger trees (higher L) to have greater *diversification potential*, i.e., the scope of the stock market to eliminate specific risk of the minimum risk portfolio. In order to look at this, we calculated the mean-variance frontiers for the ensemble of 477 stocks using $T = 1000$ as the window width. If we study the level of portfolio risk as a function of time, we find a similarity between the risk curve and the curves of the mean correlation coefficient $\bar{\rho}$ and normalized tree length L [6]. Earlier, when the smaller dataset of 116 stocks - consisting primarily important industry giants - was used, we found Pearson's linear correlation between the risk and the mean correlation coefficient $\bar{\rho}(t)$ to be 0.82, while that between the risk and the normalized tree length $L(t)$ was -0.90. Therefore, for that dataset, the normalized tree length was able to explain the diversification potential of the market better than the mean correlation coefficient. For the current set of 477 stocks, which includes also less influential companies, the Pearson's linear and Spearman's rank-order correlation coefficients between the risk and the mean correlation coefficient are 0.86 and 0.77, and those between the risk and the normalized tree length are -0.78 and -0.65, respectively.

Thus far, we have only examined the location of stocks in the minimum risk portfolio, for which $\theta = 0$. However, we note that as we increase θ towards unity, portfolio risk as a function of time soon starts behaving very differently from the mean correlation coefficient and normalized tree length as shown in Fig. 6. Consequently, it is no longer useful in describing diversification potential of the market. However, another interesting result is noteworthy: The

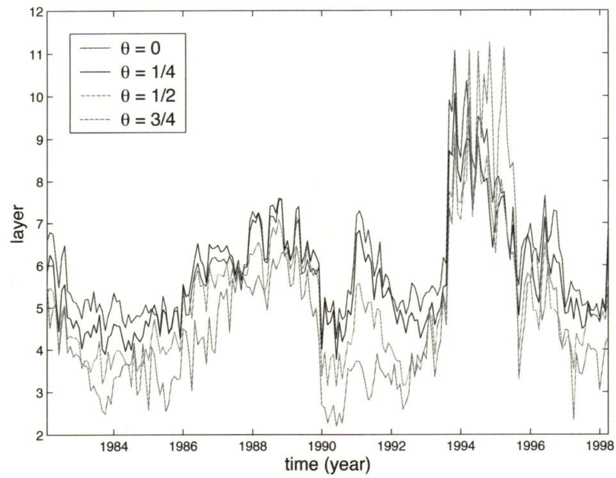

Fig. 6. Plots of the weighted minimum risk portfolio layer $l_{\mathbf{P}}(t, \theta)$ for different values of θ.

average weighted portfolio layer $l_{\mathbf{P}}(t, \theta)$ decreases for increasing values of θ. This implies that out of all the possible Markowitz portfolios, the minimum risk portfolio stocks are located furthest away from the central vertex, and as we move towards portfolios with higher expected return, the stocks included in these portfolios are located closer to the central vertex. It may be mentioned that we have not included the weighted portfolio layer for $\theta = 1$, as it is not very informative. This is due to the fact that the maximum return portfolio comprises only one asset (the maximum return asset in the current time window) and, therefore, $l_{\mathbf{P}}(t, \theta = 1)$ fluctuates wildly as the maximum return asset changes over time.

We believe these results to have potential for practical application. Stocks included in low risk portfolios are consistently located further away from the central node than those included in high risk portfolios. Consequently, the radial distance of a node, i.e. its occupation layer, is meaningful. We conjecture that the location of a company *within* the cluster reflects its position with regard to internal, or cluster specific, risk. Thus the characterization of stocks by their branch, as well as their location within the branch, would enable us to identify the degree of interchangeability of different stocks in the portfolio. In most cases, we would be able to pick two stocks from different asset tree clusters, but from nearby layers, and interchange them in the portfolio without considerably altering the characteristics of the portfolio. Therefore, dynamic asset trees may facilitate *incorporation of subjective judgment* in the portfolio optimization problem.

3 Summary

We have studied the dynamics of asset trees and applied it to market taxonomy and portfolio analysis. We have noted that the tree evolves over time and the mean occupation layer fluctuates as a function of time, and experiences a downfall at the time of market crisis due to topological changes in the asset tree. For the portfolio analysis, it was found that the stocks included in the minimum risk portfolio tend to lie on the outskirts of the asset tree: on average the weighted portfolio layer can be almost one and a half levels higher, or further away from the central vertex, than the mean occupation layer for window width of four years. Finally, the asset tree can be used as a visualization tool, and even though it is strongly pruned, it still retains all the essential information of the market (starting from the correlations in stock prices) and can be used to add subjective judgement to the portfolio optimization problem.

Acknowledgements

The author would like to thank all his collaborators, and also the critics for their valuable comments during the lectures given at IPST (Maryland, USA), Bose Institute (Kolkata) and MMV (BHU, Varanasi).

References

1. Bachelier L (1900) Theorie de la Speculation. Gauthier-Villars, Paris
2. Chakraborti A, Santhanam MS (2005) Int. J. Mod. Physics 16:1733
3. Tsay RS (2002) Analysis of Financial Time Series John Wiley, New York
4. Mantegna RN, Stanley HE (2000) An Introduction to Econophysics, Cambridge University Press, New York
5. Onnela J-P (2002) Taxonomy of Financial Assets, M. Sc. Thesis, Helsinki University of Technology, Finland
6. Onnela J-P, Chakraborti A, Kaski K, Kertesz J (2003) Phys. Rev. E 68:056110
7. Mantegna RN (1999) Eur. Phys. J. B 11:193
8. Onnela J-P, Chakraborti A, Kaski K, Kertesz J (2002) Physica A 324:247
9. http://www.forbes.com/ (2003)
10. Onnela J-P, Chakraborti A, Kaski K, Kertesz J (2002) Eur. Phys. J. B 30:285
11. http://www.lce.hut.fi/~jonnela/

The Power (Law) of Indian Markets: Analysing NSE and BSE Trading Statistics

Sitabhra Sinha and Raj Kumar Pan

The Institute of Mathematical Sciences, C. I. T. Campus, Taramani, Chennai - 600 113, India. sitabhra@imsc.res.in

The nature of fluctuations in the Indian financial market is analyzed in this paper. We have looked at the price returns of individual stocks, with tick-by-tick data from the National Stock Exchange (NSE) and daily closing price data from both NSE and the Bombay Stock Exchange (BSE), the two largest exchanges in India. We find that the price returns in Indian markets follow a fat-tailed cumulative distribution, consistent with a power law having exponent $\alpha \sim 3$, similar to that observed in developed markets. However, the distributions of trading volume and the number of trades have a different nature than that seen in the New York Stock Exchange (NYSE). Further, the price movement of different stocks are highly correlated in Indian markets.

1 Introduction

Over the past decade, a growing number of physicists have got involved in searching for statistical regularities in the behavior of financial markets. A key motivation for such "econophysicists" is the prospect of discovering *universal* features in financial data [1], i.e., statistical properties that are invariant with respect to stocks, markets, the time interval over which the data is collected, etc. The most prominent candidate for such universality is the distribution of fluctuations in the price of individual stocks [2, 3], as well as, market indices [4] which reflect the composite value of many such stocks. Studies in various markets have reported evidence for the cumulative distribution of price fluctuations having positive and negative tails that obey a power law decay, i.e., $P_c(x) \sim x^{-\alpha}$. It has also been claimed that the exponent for this power law, α, is around 3 for most markets (the "inverse cubic law") [5]. It may be useful here to distinguish between the power law reported for individual stock price fluctuations and that for market index fluctuations, as the former is more fundamental and implies the latter, provided most of the stocks comprising the index have significant cross-correlation in their price movement.

We will, therefore, focus on the behavior of individual stocks, although we will also mention in brief our study of a particular Indian market index.

The prime motivation for our study of the Indian financial market is to check recent claims that emerging markets (including those in India) have behavior that departs significantly from the previously mentioned "universal" behavior for developed markets. Although a recent paper [6] reported heavy tailed behavior of the fluctuation distribution for an Indian market index between Nov 1994 and Oct 2004, the generalized Pareto distribution fit to the data did not suggest a power law decay of the tails. Moreover, an almost contemporaneous study [7] of the fluctuations in the price of 49 largest stocks in the NSE between Nov 1994 and Jun 2002, has claimed that the distribution has exponentially decaying tails. This implies the presence of a characteristic scale, and therefore, the breakdown of universality of the power law tail for the price fluctuation distribution. The contradiction between the results of the two groups indicates that a careful analysis of the Indian market is necessary to come to a conclusive decision. Note that, both of the above-mentioned studies looked at low-resolution data, namely, the daily closing time series.

In this study, we have looked at the high-frequency transaction by transaction stock price data, as well as taken a fresh look at the low-frequency daily data. We conclude that, far from being different, the distribution of price fluctuations in Indian markets is *quantitatively* almost identical to that of developed markets. However, the distributions for trading volume and number of trades seem to be market-specific, with the Indian data being consistent with a log-normal distribution for both of these quantities. Next, we look at the distribution of fluctuations in the NSE market index and find it to also follow the "inverse cubic law". Given the result for the price fluctuation distribution of individual stocks, this is expected if the price movements of the various stocks are highly correlated. Therefore, we also study the cross-correlations among the price fluctuations of most of the stocks comprising the index. We find that, on the whole, stock price movements in the Indian market are remarkably correlated.

2 The Indian financial market

There are 23 different stock markets in India. The two largest are the National Stock Exchange (NSE) and the Bombay Stock Exchange (BSE) which together accounted for more than 98% of the total turnover for all markets in 2003-04 [8]. Of these, the NSE is the larger one, with a turnover that is slightly more than double that of BSE, although their market capitalizations are comparable. BSE was founded in 1875, and is the oldest stock market in Asia. It has the largest number of companies listed and traded, among all the exchanges in India. The market indices associated with it, namely BSE 30, BSE 100 and BSE 500, are closely followed indicators of the health of the Indian financial market. The stocks belonging to BSE 500 represent nearly

93% of the total market capitalisation in that exchange, and therefore in this study we have confined ourselves to these stocks.

Compared to BSE, NSE is considerably younger, having commenced operations in the capital (equities) market from Nov 1994. However, as of 2004, it is already the world's third largest stock exchange (after NASDAQ and NYSE) in terms of transactions [8]. The most important market index associated with the NSE is the Nifty. The 50 stocks comprising the Nifty index represent about 58% of the total market capitalization and 47% of the traded value of all stocks in the NSE (as of Dec 2005).

Description of the data set. The low-frequency data that we analyze consists of the daily closing price, volume and number of trades for individual stocks from BSE (starting from as early as 1991) and NSE (starting from as early as 1994). This data is available from the web-sites of the corresponding exchanges [9]. The high-frequency tick-by-tick data contains information of all transactions carried out in the NSE between Jan 1, 2003 and Mar 31, 2004. This information includes the date and time of trade, the price of the stock during transaction and the volume of shares traded. This database is available in the form of CDs published by NSE. For calculating the price return, we have focused on 479 stocks, which were all used to calculate the BSE 500 index during this period. To calculate the distribution of index fluctuations, we have looked at the daily closing value of Nifty between Jan 1, 1995 and Dec 31, 2005. For cross-correlation analysis, we have focused on daily closing price data of 45 NSE stocks (all belonging to the Nifty index) from Jan 1, 1997 to Dec 31, 2005.

3 Price return distribution of individual stocks

To measure the price fluctuations (or the fluctuations in the market index) such that the result is independent of the scale of measurement, we calculate the logarithmic return of price (or index). If $P_i(t)$ is the stock price of the ith stock at time t, then the (logarithmic) price return is defined as

$$R_i(t, \Delta t) \equiv \ln P_i(t + \Delta t) - \ln P_i(t). \quad (1)$$

However, the distribution of price returns of different stocks may have different widths, owing to differences in their volatility. To be able to compare the distribution of various stocks, we must normalize the returns by dividing them with their standard deviation (which is a measure of the volatility), $\sigma_i = \sqrt{\langle R_i^2 \rangle - \langle R_i \rangle^2}$. The normalized price return is, therefore, given by

$$r_i(t, \Delta t) \equiv \frac{R_i - \langle R_i \rangle}{\sigma_i}, \quad (2)$$

where $\langle \ldots \rangle$ represents time average.

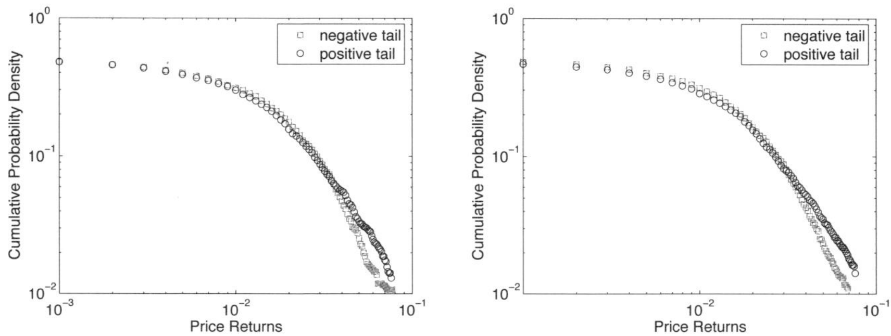

Fig. 1. Cumulative distribution of daily price returns for a particular stock (Reliance) at BSE (left) between July 12, 1995 and Jan 31, 2006, and at NSE (right) between Nov 3, 1994 and Jan 30, 2006.

3.1 Daily price returns in BSE and NSE

We start by focussing on the daily price variation of individual stocks, i.e., $\Delta t = 1$ day. By using the time series of daily closing price of a particular stock (e.g., Reliance) we can obtain the corresponding daily returns. Binning this data appropriately we can obtain the probability density function, and by integrating it over a suitable range, the cumulative distribution function (CDF), which is essentially the probability that a return is larger than a given value. Fig. 1 shows the CDF for daily price returns for the same stock in BSE (left) and NSE (right). Note that, we have shown the tails for the positive and negative returns in the same figure. The distribution for the two exchanges are almost identical, and both show long tails consistent with a power law decay.

To confirm that this is a general property, and not unique to the particular stock that is being analysed, we next perform the same analysis for other stocks. To be able to compare between stocks, we normalize the returns for each stock by their standard deviation. Fig. 2 (left) shows that four stocks chosen from different sectors have very similar normalized cumulative distributions. Moreover, the tail of each of these distributions approximately follow a power law with exponent $\alpha \simeq 3$. However, the daily closing price data set for any particular stock that we have analyzed is not large enough for an unambiguous determination of the nature of the tail. For this, we aggregate the data for 43 frequently traded stocks, all of which are used for calculating the Nifty index, over 3 years, and obtain the corresponding CDF (Fig. 2, right). Putting together the time series of different stocks to form a single large time series is justified because, after normalization, the different stocks have almost identical distributions [3]. From this figure we confirm that the distribution does indeed follow a power law decay, albeit with different exponents for the positive and negative return tails. The different exponents of the positive and negative tails have also been observed in the case of stocks listed in the New

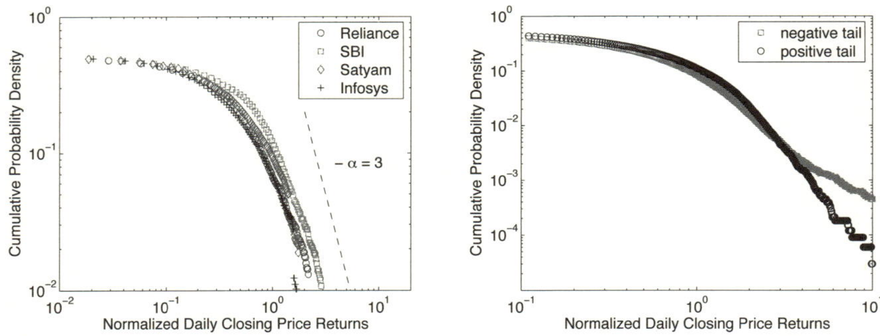

Fig. 2. (Left) Cumulative distribution of the normalized daily closing price returns for four stocks in BSE between July 12, 1995 and Jan 31, 2006. (Right) Cumulative distribution of the aggregated normalized daily closing price returns for 43 stocks (included in the Nifty index) at NSE between Jan 1, 2003 and Jan 30, 2006.

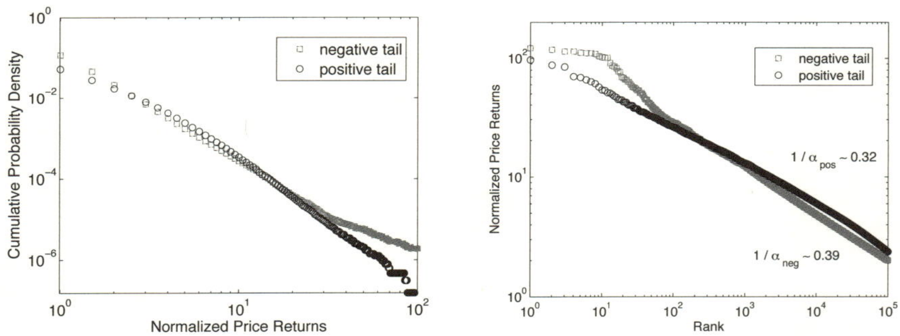

Fig. 3. Cumulative distribution (left) and rank-ordered plot (right) for the 5-minute interval price returns aggregated for 479 stocks at NSE between Jan 1, 2003 to Mar 31, 2004.

York Stock Exchange (NYSE) [3]. For comparison, we carried out a similar study with the daily closing price series of several NYSE stocks from Yahoo! Finance [10], and obtained distributions qualitatively similar to that shown here for the Indian market.

3.2 Price return for tick-by-tick data in NSE

The daily data is strongly suggestive of a power law tail for the price return distribution, but for conclusive evidence we next turn to the tick-by-tick data for stocks listed in the NSE. Choosing an appropriate Δt, we obtain the corresponding return by taking the log ratio of consecutive average prices, averaged over a time window of length Δt; for the results reported here $\Delta t = 5$ minutes. We have verified that the nature of the distribution is not sensitive to the exact value of this parameter. For individual stocks, the cumulative distribution of returns again show power law decay, but as the data set for each stock is not large enough, we carry out an aggregation procedure similar to that outlined above. Picking 479 frequently traded stocks from NSE, we put together their normalized returns to form a single large data set. The corresponding CDF is shown in Fig. 3 (left), with the exponents for the positive and negative tails estimated to be $\alpha \sim 3.2$ and 2.7, respectively. To check the accuracy of these exponents, obtained using linear least square fitting on a doubly logarithmic plot, we next plot the return data in descending order. This *rank-ordered plot* is an alternative visualization of the CDF, interchanging the ordinate and abscissae. It is easy to show that if the CDF has a power-law form, so does the rank-ordered plot, and the two exponents are the inverses of each other [11]. Exponents obtained by least square fitting on this graph produces similar values of α, namely, 3.1 and 2.6 for the positive and negative tails, respectively.

3.3 The "inverse cubic law" for price and index fluctuations

The results reported above provide conclusive evidence that the Indian financial market follows a price fluctuation distribution with long tails described by a power law. Moreover, the exponent characterizing this power law is close to 3, as has been observed for several financial markets of developed economies, most notably the NYSE, where the "inverse cubic law" has been found to be valid from $\Delta t = 1$ day to 1 month.

Most observations of this "law" have been in the context of market indices, rather than the price of individual stocks. We have, therefore, carried out a similar analysis for the Nifty index of NSE during the period Jan 1, 1995 to Dec 31, 2005. Fig. 4 (left) shows that the distribution of index returns also shows a power law decay, with an exponent very close to 3. As the index is a

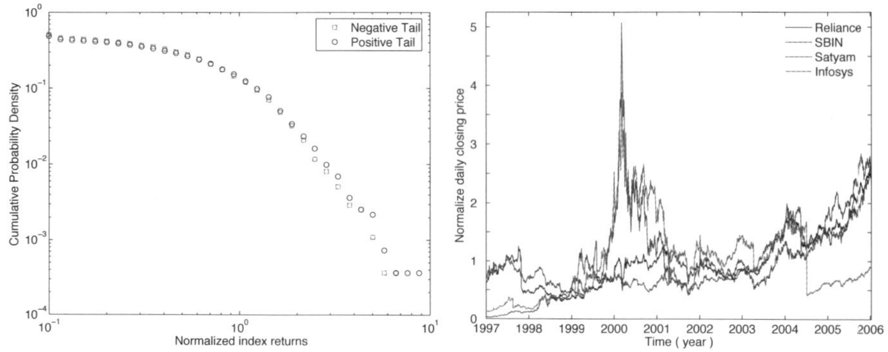

Fig. 4. (Left) Cumulative distribution of daily returns for the Nifty index between Jan 1, 1995 and Dec 31, 2005. (Right) Comparison of the daily closing price for four stocks in NSE from Jan 1, 1997 to Dec 31, 2005, showing the high degree of correlation among the stocks.

composite of several stocks, this behavior can be understood as a consequence of the power law decay for the tails of individual stock price returns, provided the movement of these stocks are correlated. As is evident from Fig 4 (right), this condition is indeed satisfied in the Indian market. In a later section we provide a more detailed look into the cross-correlation structure of these price fluctuations.

These findings assume importance in view of the recent claims that emerging markets behave very differently from developed markets, in particular, exhibiting an exponentially decaying return distribution [7]. India is one of the largest emerging markets, and our analysis of the price fluctuations in the major Indian stock exchanges challenges these claims, while at the same time, providing strong support to the universality for the "inverse cubic law" which had previously only been seen in developed markets.

4 Distribution of trading volume and number of trades

Besides the price of stocks, one can also measure market activity by looking at the trading volume (the number of shares traded), $V(t)$, and the number of trades, $N(t)$. To obtain the corresponding distributions, we normalize these variables by subtracting the mean and dividing by their standard deviation, such that, $v = \frac{V-\langle V\rangle}{\sqrt{\langle V^2\rangle-\langle V\rangle^2}}$ and $n = \frac{N-\langle N\rangle}{\sqrt{\langle N^2\rangle-\langle N\rangle^2}}$. Fig. 5 shows the distribution of these two quantities for several stocks, based on daily data for BSE. As is evident, the distribution is very similar for the different stocks, and the nature of the decay is significantly different from a power law. To better characterize the distribution, we have also looked at the intra-day distributions for volume and number of trades, based on high-frequency data from NSE. Fig. 6 shows the distributions of the two quantities for trading conducted on a particular stock in 5 minute intervals. Analysis of data for other stocks show qualitatively

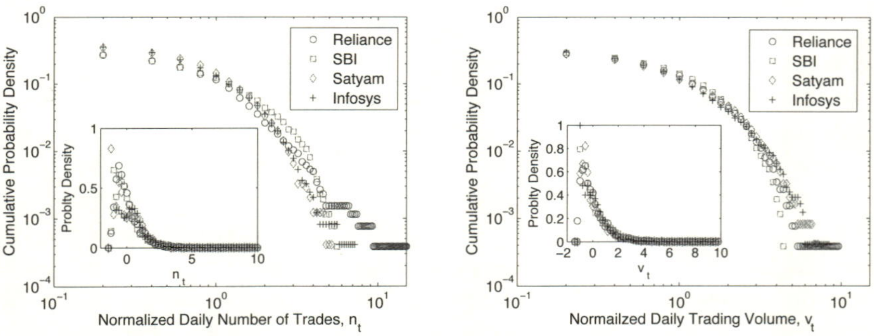

Fig. 5. Cumulative distribution of the number of trades (top left) and the volume of shares traded (top right) in a day for four stocks at BSE between July 12, 1995 and Jan 31, 2006.

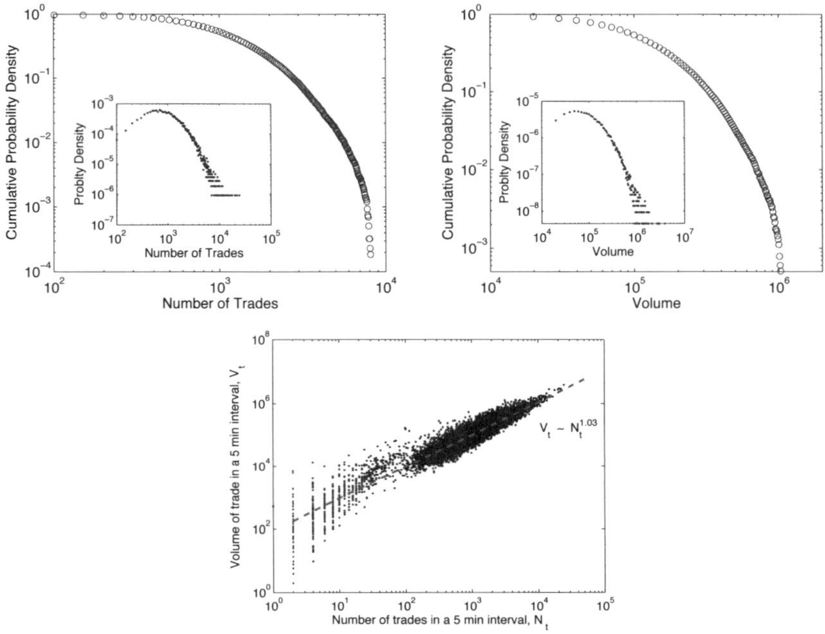

Fig. 6. Cumulative distribution of the number of trades (top left) and the volume of shares traded (top right) for a particular stock (Reliance) in 5-minute intervals at NSE between Jan 1, 2003 to March 31, 2004. The bottom figure shows an almost linear relation between the number of trades in a 5-minute interval and the corresponding trading volume. The broken line indicates the best fit on a doubly logarithmic scale.

similar results. As is clear, both of these distributions are non-monotonic, and are suggestive of a log-normal form. The fact that these distributions are very similar to each other is not surprising in view of the almost linear relationship between the two (Fig. 6, bottom). This supports previous observation in major US stock markets that statistical properties of the number of shares traded and the number of trades in a given time interval are closely related [13].

For US markets, power law tails have been reported for the distribution of both the number of trades [12] and the volume [13]. It has also been claimed that these features are observed on the Paris Bourse, and therefore, these features are as universal as the "inverse cubic law" for price returns distribution [14]. However, analysis of other markets, e.g., the London Stock Exchange [15] have failed to see any evidence of power law behavior. Our results confirm the latter assertion that the power law behavior in this case may not be universal, and the particular form of the distribution of these quantities may be market specific.

5 Correlated stock movement in the Indian market

As indicated in a previous section, we now return to look at the cross-correlation among price movements. The data that we analyze for this purpose consists of 2255 daily returns each for 45 stocks. We divide this data into M overlapping windows of width T, i.e., each window contains T daily returns. The displacement between two consecutive windows is given by the window step length parameter δt. In our study, T is taken as six months (125 trading days), while δt is taken to be one month (21 trading days). The correlation between returns for stocks i and j is calculated as

$$C_{ij} = \langle r_i r_j \rangle - \langle r_i \rangle \langle r_j \rangle, \tag{3}$$

where $\langle \ldots \rangle$ represents the time average within a window. The resulting correlation matrices, C, can be analysed to get further understanding of the relations between movements of the different stocks.

We now look at the eigenvalues of C which contain significant information about the cross-correlation structure [16]. Fig. 7 (left) shows the eigenvalues of C as a function of time. It is clear that the majority of these are very close to zero at all times. The largest eigenvalues contain almost all information about the market, which is evident from Fig. 7 (right). This shows the variation of the average correlation coefficient, as well as the largest eigenvalue λ_{max}, with time. The two are strongly correlated, indicating that λ_{max} captures the behavior of the entire market. Our results indicate that the Indian market is highly correlated, as indicated by the strong cross-correlations among the most traded stocks.

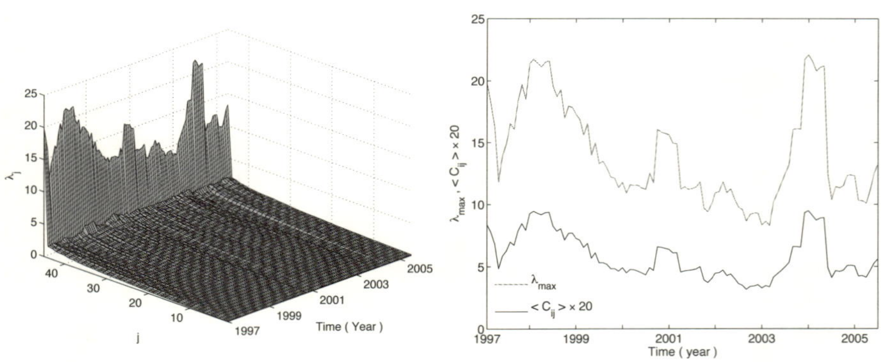

Fig. 7. (Left) The eigenvalues, sorted in descending order, for the correlation matrices of daily price returns for 45 stocks in NSE, across time. (Right) The variation across time of the largest eigenvalue λ_{max} of the correlation matrices and the average correlation $\langle C \rangle$. The window size chosen for calculating correlation is 6 months and the window is shifted in steps of 21 days. The total period is from Jan 1997 to Dec 2005.

6 Conclusions

In this paper, we have examined the statistical properties of trading in the two largest Indian financial markets, BSE and NSE. Using both low-frequency (daily) and high-frequency (tick-by-tick), we demonstrate that the price return cumulative distribution has long tails, consistent with a power law having exponent close to 3. This lends strong support to the claim that the price return distribution has an universal form across different markets, namely, the "inverse cubic law". On the other hand, the distributions for volume and number of trades appear to be log-normal, the two quantities being almost linearly related. We also find the market index fluctuation distribution to have the same form as the distribution of individual stock price returns. This implies that stocks in the Indian market are highly correlated. We verify that this is indeed the case with a cross-correlation analysis of most of the frequently traded stocks in the Indian market.

Acknowledgements

We are grateful to M. Krishna for invaluable assistance in obtaining and analyzing the high-frequency NSE data. We thank S. Sridhar and N. Vishwanathan for technical assistance in arranging the data, and J.-P. Onnela for helpful discussions.

References

1. Farmer JD, Shubik M, Smith E (2005) Physics Today 58(9): 37–42
2. Lux T (1996) Applied Financial Economics 6: 463–475
3. Plerou V, Gopikrishnan P, Amaral LAN, Meyer M, Stanley HE (1999) Phys. Rev. E 60:6519–6529
4. Gopikrishnan P, Plerou V, Amaral LAN, Meyer M, Stanley HE (1999) Phys. Rev. E 60: 5305–5316
5. Gopikrishnan P, Meyer M, Amaral LAN, Stanley HE (1998) Eur. Phys. J. B 3: 139–140
6. Sarma M (2005) EURANDOM Report 2005-003 (http://www.eurandom.tue.nl/reports/2005/003MSreport.pdf)
7. Matia K, Pal M, Salunkay H, Stanley HE (2004) Europhys. Lett. 66: 909–914
8. National Stock Exchange (2004) Indian securities market: A review. (http://www.nseindia.com/content/us/ismr2005.zip)
9. BSE: http://www.bseindia.com/, NSE: http://www.nseindia.com/
10. http://finance.yahoo.com/
11. Adamic LA, Huberman BA (2002) Glottometrics 3:143–150
12. Plerou V, Gopikrishnan P, Amaral LAN, Gabaix X, Stanley HE (2000) Phys. Rev. E 62: 3023–3026
13. Gopikrishnan P, Plerou V, Gabaix X, Stanley HE (2000) Phys. Rev. E 62: 4493–4496

14. Gabaix X, Gopikrishnan P, Plerou V, Stanley HE (2003) Nature 423: 267–270
15. Farmer JD, Lillo F (2004) Quantitative Finance 4: C7–C11
16. Plerou V, Gopikrishnan P, Rosenow B, Amaral LAN, Guhr T, Stanley HE (2002) Phys. Rev. E 65: 066126

A Random Matrix Approach To Volatility In An Indian Financial Market

V. Kulkarni and N. Deo

Department of Physics and Astrophysics, University of Delhi, Delhi 110007, India
ndeo@physics.du.ac.in, varsha_s_kulkarni@yahoo.com

Summary. The volatility of an Indian stock market is examined in terms of aspects like participation, synchronization of stocks and quantification of volatility using the random matrix approach. Volatility pattern of the market is found using the Bombay Stock Index for the three-year period 2000 − 2002. Random matrix analysis is carried out using daily returns of 70 stocks for several time windows of 85 days in 2001 to (i) do a brief comparative analysis with statistics of eigenvalues and eigenvectors of the matrix C of correlations between price fluctuations, in time regimes of different volatilities. While a bulk of eigenvalues falls within Random Matrix Theory bounds in all the time periods, we see that the largest (deviating) eigenvalue correlates well with the volatility of the index (ii) observe the corresponding eigenvector clearly shows a shift in the distribution of its components from volatile to less volatile periods and verifies the qualitative association between participation and volatility (iii) set up a variability index, V whose temporal evolution is found to be significantly correlated with the volatility of the overall market index.

1 Introduction

Physical phenomena (like Brownian motion [1], turbulence, chaos [2]) have found application in the study of the dynamics of financial markets. Financial time series originate from complex dynamical processes, sometimes accompanied by strong interactions. The nature of underlying interactions in a stock market is not known much the same as in complex quantum systems. A number of researchers [4–11] have applied the methods of RMT [3] to financial data and found interesting clues about the underlying interactions. This paper is a modest attempt, to exposit some observations that may throw light on volatility (market fluctuations). The purpose of this paper is two-fold. First, it attempts to understand quantitatively the closely related aspects of volatility such as synchronization and participation of stocks in the market using the random matrix technique and second, to show that this technique may be used to set up a quantity which possesses a strong predictive power for the volatility of the market.

We start with a brief empirical analysis of the BSE index and show the volatility pattern. The next section deals with the random matrix approach and the financial correlation matrix. We conclude by discussing our observations in the last section.

2 Empirical analysis of BSE index

We first consider statistical properties of the time evolution of BSE index. We label the time series of index as $Y(t)$. We calculate volatility and find the p.d.f of index changes.

2.1 Data analyzed:

This section uses the daily indices of the Bombay Stock Exchange (BSE) for a period of 3 years between 2000 – 2002. Indices are basically averages of actively traded stocks, which are weighted according to their market value. Trading is done five days a week in this market, each year corresponds roughly to 250 days of elapsed time, the total number of data points in this set is 750.

2.2 Volatility

Volatility is a measure of fluctuations that occur in the market. Statistical properties of volatility prove to be of vital practical significance as volatility is a key parameter in risk management. It is directly linked to the information arriving in speculative markets. Rare events or unanticipated 'shocks', seasonal changes, economic and political cycles (elections, the announcement of the budget in a country) all tend to accentuate fluctuations. A highly volatile period is marked by increased uncertainty and a greater propensity of traders to speculate and interact.

Computing Volatility: Volatility, as mentioned earlier, gives us a measure of the market fluctuations. Intuitively we can say that a stock whose prices fluctuate more is more "risky" or "volatile". We may formalize this as : Let $Y(t-\Delta t), Y(t), Y(t+\Delta t), \cdots$ be a stochastic process where $Y(t)$ may represent prices, indices, exchange rates etc. The logarithmic returns $G(t)$ over time scale Δt are

$$
\begin{aligned}
G(t) &= \log(Y(t+\Delta t)) - \log(Y(t)) \\
&\approx \frac{Y(t+\Delta t)}{Y(t)} \text{in the limit of small changes to Y.}
\end{aligned}
\tag{1}
$$

Δt refers to the time interval. In this case $\Delta t = 1$ day.

We quantify volatility, as the local average of the absolute value of daily returns of indices in an appropriate time window of T days, as an estimate of volatility in that period, see reference [12],

$$v(t) = \frac{\sum_{t=1}^{T-1} |G(t)|}{T-1} \tag{2}$$

We compute monthly volatility for the three year period 2000 – 2002 by taking $T = 20$ days as there are roughly 20 trading days in a month in BSE. $\sum |G(t)|$ may be considered as a substitute for volatility or 'scaled volatility' in future.

The BSE index for the period 2000 – 2002, is shown in figure 1. The figure shows a significant change in the value of index over the period of three years (2000 – 2002). The rate of change (decrease) appears to be more for the first 450 days than later. We may say that the Bombay stock exchange follows a long-term trend in the period considered in the sense that there is more uncertainty say four months in future than a month in future. The trend also reflects on the willingness to take risk on part of the traders; it seems the market was far more active in the year 2000 than 2001 or 2002. There is a sharp dip in Y near 9/11/2001 (425th day) after which the index rises and settles without much fluctuation. This is indicated in Figure 2.

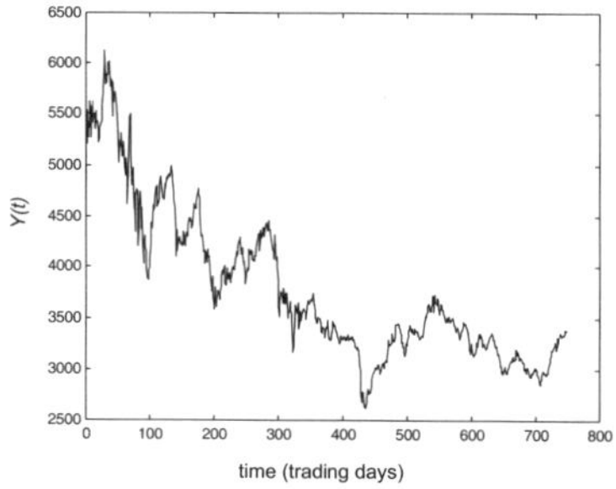

Fig. 1. BSE index plotted for all the days reported in the period 2000–2002. Total number of points plotted is 750. A sharp dip can be seen around September 11th, 2001 (425th day in figure) when the index drops to the lowest.

Figure 3 shows the volatility of the market in the period 2000-2002. It is interesting to see from here three periods characterized by distinct volatilities respectively. Each year corresponds to 250 days of elapsed time and we may divide the period into three sub periods, I- 1 – 250 days, II-251 – 500 days, III- 501 – 750 days with scaled volatilities 5.65, 3.5 and 2.25 respectively. We see that the year 2000 (1 – 250 days) was extremely active showing very high fluctuations in the market and that regions of high volatility occur in clusters showing consistently high fluctuations in say the first 200 days and more. Subsequently the fluctuations decrease in 2001 (251 – 500 days) . The period marked 420 – 440 shows a drastic jump indicating that the event of

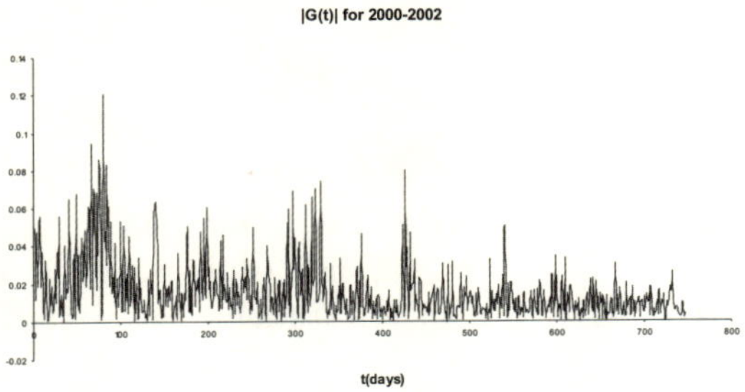

Fig. 2. Plot of absolute value of logarithmic returns of indices $|G|$ for the period 2000–2002. Large values of $|G|$ indicate crashes such as one on September 11th 2001 (425th day in the figure).

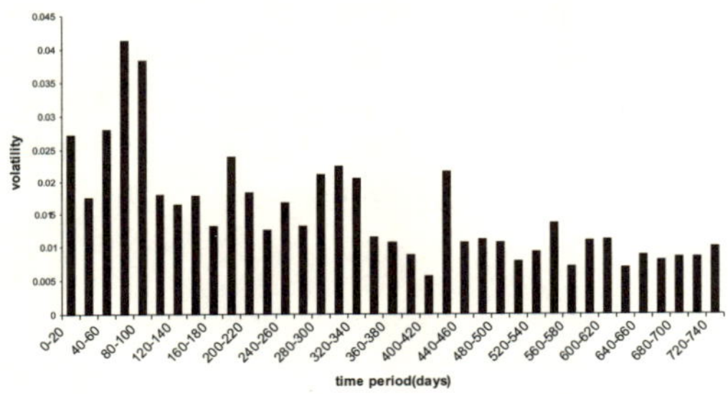

Fig. 3. Volatility,v of BSE index for 20-day periods between 2000-2002. Last ten days of the period have been ignored. The period 420-440 days including the date of September 11th 2001 shows a sudden burst of activity.

9/11 which is represented as the 425th day in the set, increased the volatility for that period. This sudden burst does not show any precursors, as it was an unanticipated event that rattled the market in that period. The impact of the event of 9/11 was not long lasting, the last time period shows very little fluctuation indicating a quiescent state in 2002 (501 days onwards).

Since volatility is measured as the magnitude of the index changes, it may be worthwhile here to compare the return distributions for the three time periods as discussed above.

We calculate the probability distribution function (p.d.f), $P(Z)$ of daily returns, Z

$$Z_{\Delta t} = Y(t + \Delta t) - Y(t); \qquad \Delta t = 1 \text{ day} \tag{3}$$

The three distributions (Figure 4) differ in mean and standard deviation (σ). While the periods: II and III do not differ much in maximum probability, the figure clearly shows the value of probability for small index changes, $P(Z - \delta Z < Z < Z + \delta Z)$ for the period I is significantly lesser than (less than half) the corresponding values in the other two periods. However period I shows a fatter tail, that is, it shows a higher probability for larger index changes than II, III.

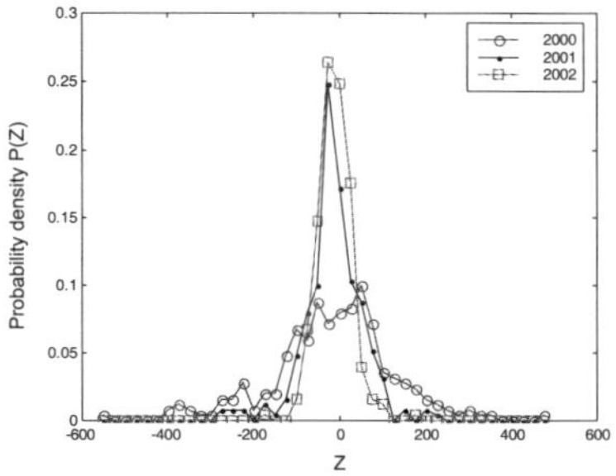

Fig. 4. The probability distributions of daily index changes Z are shown for three periods I, II, III corresponding to the years 2000, 2001, 2002 having mean $= -4.88$, $\sigma = 141.58$, for period I; mean=-2.76, $\sigma = 69.96$, for period II; mean=0.53, $\sigma = 39.01$, for period III. The value of δz is taken as 25.

3 Random matrix approach

Random Matrix Theory was developed by Wigner, Dyson and Mehta [3] in order to study the interactions in complex quantum systems. It was used to explain the statistics of energy levels therein. RMT has been useful in the analysis of universal and non-universal properties of cross-correlations between different stocks. Recently various studies [4]– [11] have quantified correlations between different stocks by applying concepts and methods of RMT, and have shown that deviations of properties of correlation matrix of price fluctuations of stocks, from a random correlation matrix yield true information about the existing correlations. While the deviations have been observed and studied in detail in the context of financial markets in earlier studies, we make a comparative analysis here, in the context of volatile versus less volatile situations from the point of view of correlations, participation of stocks in the market and try to quantify volatility in terms of the deviations.

3.1 Data analyzed and constraints involved:

Many of the stocks in BSE are not actively traded and hence not reported regularly in any period of time. Consequently they do not contribute much to the variations in stock price indices. Hence we consider here seventy actively traded stocks from largest sectors like chemical industry, metal and non-metal (diversified including steel, aluminum, cement etc). Periods of analysis are confined to $280 - 500$ days.

3.2 Cross correlations

We quantify correlations for T observations of inter day price changes (returns) of every stock $i = 1, 2, \cdots, N$ as

$$G_i(t) = \log P_i(t+1) - \log P_i(t) \tag{4}$$

where $P_i(t)$ denotes the price of stock $i = 1, 2, \cdots, N$ and $t = 1, 2, \cdots, T-1$. Since different stocks vary on different scales, we normalize the returns as

$$M_i(t) = \frac{G_i(t) - < G_i >}{\sigma} \tag{5}$$

where $\sigma = \sqrt{< G_i^2 > - < G_i >^2}$ is the standard deviation of G_i. Then the cross correlation matrix C, measuring the correlations of N stocks is constructed with elements

$$C_{ij} = < M_i(t) M_j(t) > \tag{6}$$

The elements of C are $-1 \leq C_{ij} \leq 1$.
$C_{ij} = 1$ corresponds to complete correlation
$C_{ij} = 0$ corresponds to no correlation
$C_{ij} = -1$ corresponds to complete anti correlation.

We construct and compare the cross correlation matrix C from daily returns of $N = 70$ stocks for two analysis periods of 85 days each (i) $280 - 365$ days and (ii) $340 - 425$ (see fig 3) marked with distinct index volatilities respectively. The probability densities of elements of C, $P(C_{ij})$ for both periods are compared in Figure 5. The distribution (i) is characterized by a more positive mean. Notice that the tail in Figure 5 for positive C_{ij} starts rising again. The figure also suggests that there is a de-concentration in higher levels of correlation magnitudes in a less volatile period (ii) as compared to more volatile period (i). A clear picture of existence of more pronounced correlations in periods of high volatility is shown in Figure 8. The simple correlation coefficient between the $< |C| >$ and volatility is found to be as high as 0.94.

The eigenvalues of C have special implications in identifying the true nature of the correlations. Earlier studies using RMT methods have analyzed C and shown that 98% of eigenvalues of C lie within the RMT limits whereas 2% of them lie outside [9]. It is understood that the largest eigenvalue deviating from RMT prediction reflects that some influence of the full market is common to all stocks, and that it alone yields "genuine" information hidden in C. The second to largest eigenvalue, third to largest eigenvalue etc also carry valuable information regarding the market sectors, here we will not discuss the implications of their movements and leave it to a future publication. The range of eigenvalues within the RMT bounds corresponds to noise and do not yield any system specific information.

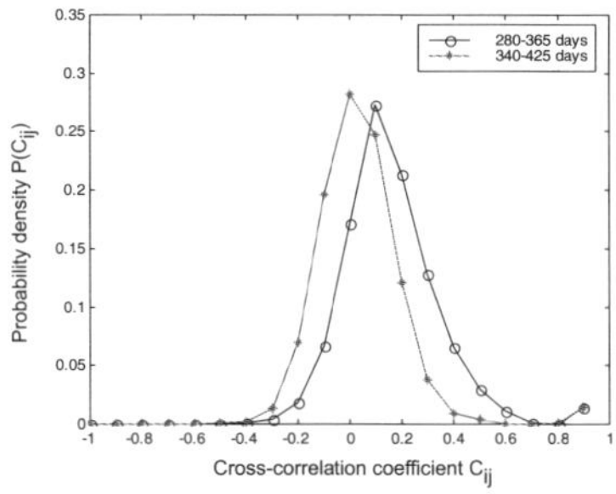

Fig. 5. Plot of the probability density of elements of correlation matrix C calculated using daily returns of 70 stocks two 85 day analysis periods (i) 280-365 days and (ii) 340-425 days with scaled volatilites of 1.6 and 0.82 respectively. We find a large value of average magnitude of correlation $<|C|>= 0.22$ for (i) compared to $<|C|>= 0.14$ for (ii).

Eigenvalue distribution of the correlation matrix

In order to extract information about the cross correlations from the matrix C, we compare the properties of C with those of a random correlation matrix. C is $N \times N$ matrix defined as

$$C = \frac{GG^\top}{T} \tag{7}$$

where G is an $N \times T$ matrix, N stocks taken for T days and $G^\top$ denotes transpose of matrix G. We now consider a random correlation matrix

$$R = \frac{AA^\top}{T} \tag{8}$$

where A is $N \times T$ matrix with random entries (zero mean and unit variance) that are mutually uncorrelated. Statistics of random matrices such as R are known. In the limit of both N and T tending to infinity, such that $Q = T/N(> 1)$ is fixed, it has been shown that the probability density function $\rho_{rm}(\lambda)$ of eigenvalues of R is given by

$$\rho_{rm}(\lambda) = \frac{Q\sqrt{(\lambda_+ - \lambda)(\lambda - \lambda_-)}}{2\pi\lambda} \tag{9}$$

for λ lying in $\lambda_- < \lambda < \lambda_+$ where λ_- and λ_+ are the minimum and maximum eigenvalues of R, respectively given by

$$\lambda_\pm = 1 + \frac{1}{Q} \pm 2\sqrt{\frac{1}{Q}} \tag{10}$$

We set up a correlation matrix C from the daily returns of $N = 70$ stocks for $T = 85$ days in the year 2001 for two periods (i) first excluding the data reported on the day -September 11, 2001 - the 85th day being Aug 31st, and then (ii) sliding this window of 85 days to include the data reported on that day and beyond - the 85th day being September 18th. Here $Q = 1.21$, and maximum and minimum eigenvalues predicted by RMT from (11) are 0.0086 and 3.6385.

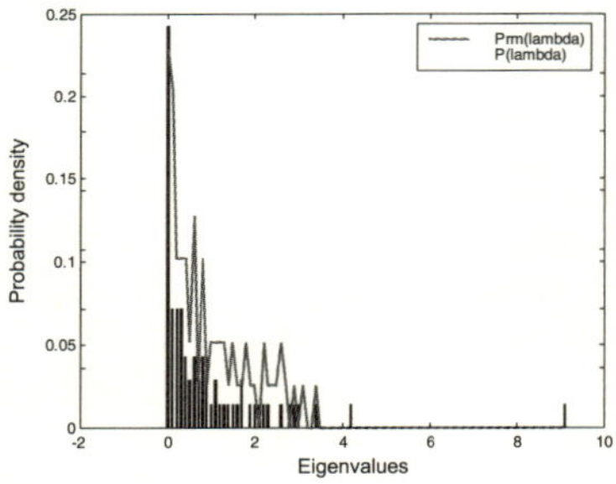

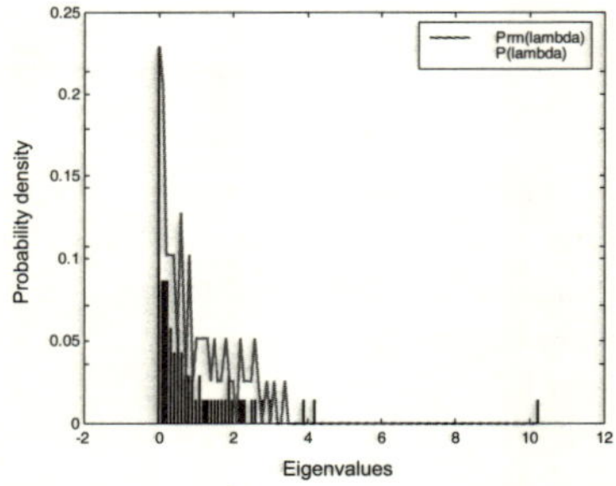

Fig. 6. Probability density of eigenvalues is shown by bars for a period considered (i) 334-419 before 9/11/2001 and having a volatility (scaled) of 0.8 (Top) and (ii) 346-431 including 9/11/2001 and having a volatility (scaled) of 0.9 (Bottom). A comparison is made with the probability density of eigenvalues of a random matrix R of the same size as C, shown by the solid line. The number of deviating eigenvalues is 4 in (i) and 6 in (ii). λ_+ for (i) is 9.17 and for (ii) is 10.28.

Trend of largest eigenvalues

Since the largest eigenvalue is believed to represent true information about the correlations between stocks and is indicative of an influence common to all stocks, we wish to see the variation of the same as we move from a no-shock, quiescent period to the one hit by the shock of 9/11. Here we start by setting up C using daily returns of $N = 70$ stocks for progressing time periods of length $T = 85$ days. We look at the largest eigenvalue of correlation matrix C. The trace of the C is preserved throughout, $Tr(C) = N$. The closer the maximum eigenvalue is to the trace, more information it contains and more correlated the prices would be. Variation of largest eigenvalue is seen by advancing the time window each time by two days. Labelling the first and last day of all periods as t_f and t_l respectively, we set up C as

$$C(t_f, t_l) = C(280 + j, 280 + j + 85) \tag{11}$$

where $j = 0, 2, 4, 6, \cdots, 134$.

Result of this exercise is shown in Figure 8. The largest eigenvalue is found to be strongly correlated with volatility of the BSE index (simple correlation coefficient is found to be 0.94) for all times considered. We study the impact of 9/11 shock by carrying out a similar exercise, taking $j = 0, 1, 2, 3, \cdots, 26$. The aftermath of the event can be seen in the sudden, impulsive rise in the maximum eigenvalue around September 13th, 18th, indicating that the impact was localized in time.

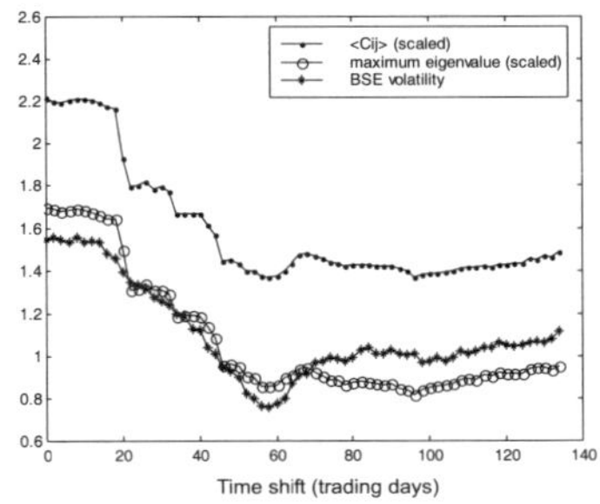

Fig. 7. Variation of largest eigenvalue and $< |C| >$, with the time shift, j. Analysis period is confined to period II. First j was increased in steps of 2 days each time to span a total time of $280 - 500$ days (see Figure 3). Volatility has been scaled for convenience. A minute exercise was carried out in Figure 8 by advancing the time windows in steps of 1 day each time, spanning a total time of $333 - 444$ days in order to capture the impact of the 9/11 shock. The horizontal axis shows the last day of all the time periods.

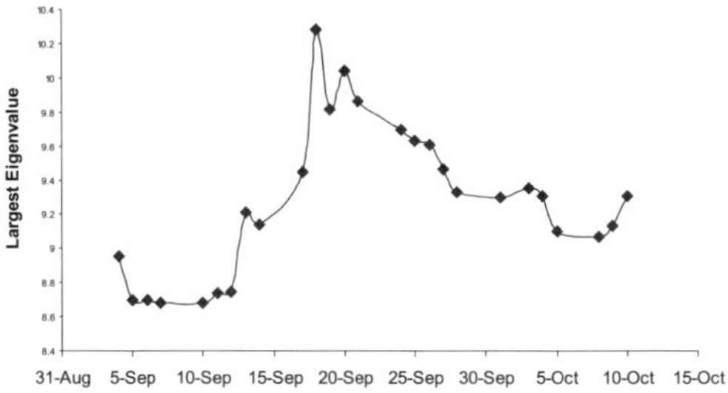

Fig. 8. Variation of largest eigenvalue with the time shift, j. Analysis period is confined to period II. First j was increased in steps of 1 days each time to span a total time of $333 - 444$ days in order to capture the impact of the 9/11 shock. (see Figure 3). Volatility has been scaled for convenience. The horizontal axis shows the last day of all the time periods.

3.3 'Last' eigenmode and the variability index

The eigenstates of C deviating from RMT predictions bring out the collective response of the market to perturbations. Collective motion of all the assets in the portfolio is significantly high, or the stocks are highly correlated in regimes marked by occasional or persisting bursts of activity. The degree of such synchronization is indicated by the eigenvector corresponding to the largest eigenvalue, through the evolution of its structure and components, seen in a. Finally in b, we try to quantify volatility in terms of the largest eigenvector to yield a strong indicator of variability.

Distribution of eigenvector components

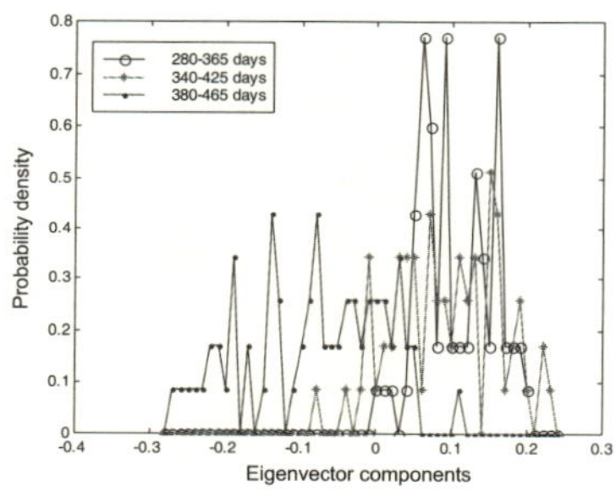

Fig. 9. Probability density of the eigenvector components for largest eigenvalue for three periods (i) 280-365 days (ii) 340-425 and (iii) 380-465 days marked by volatilities 1.6, 0.82, 0.99 respectively. The plots are for C constructed from daily returns of 70 stocks for $T = 85$ days.

We wish to analyze the distribution of the components of the eigenvector corresponding to largest eigenvalue and compare the distributions for three time periods characterized by different volatilities (i) $280 - 365$ (ii) $340 - 425$ (iii) $380 - 465$. Figure 9 shows the distributions of components of U^{70} are shorter and broader in less volatile regimes (ii,iii) than in a more volatile one (i). Although the maximum participation is more in distributions (ii), (iii) the number of significant participants i.e. components differing significantly from zero in their sets is far lesser than (almost half) that in (i). In addition we find that all the components of U^{70} for period (i) have a positive sign, which confines the distribution to one side. This finding has been interpreted previously [9] to imply that a common component of the significant participants of U^{70} affects all of them alike. We also find that the distributions for all time periods that follow (iii) and are relatively quiescent (not shown here), contain both positive and negative elements. This goes to show an interesting link between the strength of the 'common influence' and volatility. We may say 'collective' or 'ensemble-like' behavior is more pertinent to volatile situations rather than non-volatile ones.

Variability index

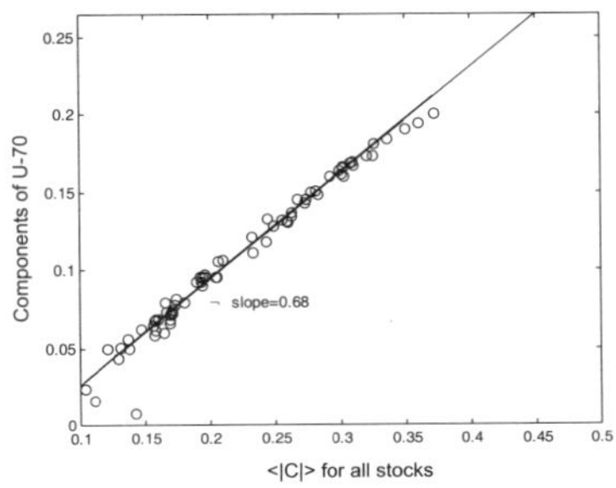

Fig. 10. Plot of the components of the eigenvector corresponding to the largest eigenvalue with the extent to which every individual stock is correlated in the market, denoted by $< |C| >_m$. In this case, correlation matrix, C was constructed using daily returns of 70 stocks for the period $280 - 365$ days. The line obtained least square fitting has a slope$= 0.68 \pm 0.01$.

A yet another interesting feature brought out in the analysis of eigenvectors is the large-scale correlated movements associated with the last eigenvector, the one corresponding to largest eigenvalue. For this purpose, the average magnitude of correlations of prices of every stock m with all stocks $n = 1, 2, \cdots, N$ is $< |C| >_m$ for $m = 1, 2, \cdots, N$ is varied with the corresponding components of U^{70} (deviating) and U^2 (lying within the bulk). While we find a strong linear positive relationship (shown in Figure 10) between the two at all times for the U^{70}, the eigenvector belonging to the RMT range (Figure 11) shows almost zero dependence. In this final sub-section we make use of this dependence to set up a Variability Index, which is strongly correlated with the variability of BSE index. We define a projection vector S with elements $< |C| >_m$ where $m = 1, 2, \cdots, 70$ as calculated before. We obtain a quantity $X_m(t)$ by multiplying each element S_m by the magnitude of the corresponding component of U_{70} for each time window 't'.

$$X_m(t) = (U_m^{70})^2 S_m, \qquad m = 1, 2, \cdots, 70 \tag{12}$$

The idea is to weight the average correlation possessed by every stock m in the market according to the contribution of the corresponding component to the last eigenvector U^70, thereby neglecting the contribution of non-significant participants (close to zero) in U^{70}. The quantity X in some sense represents the 'true' or 'effective' magnitude of correlations of stocks and the sum of such magnitudes are obtained as

$$V(t) = \sum_{m=1}^{70} X_m(t), \qquad \text{at time t} \tag{13}$$

may be expected to reflect the variability of the market at that time. We call it the Variability Index. We note from Figure 12 that the variability index behaves remarkably similarly to the volatility of BSE index as the time window is slid forward. A highly statistically significant coefficient of correlation of 0.95 is obtained and a positive, linear relationship between the two can be seen in the plot of V and BSE index volatility set out in Figure 13. We thus find the relevance of the last eigenmode in quantifying the volatility of the overall market. Similar procedures have been carried out in other studies [2] in different contexts to verify the relevance of this last eigenvector.

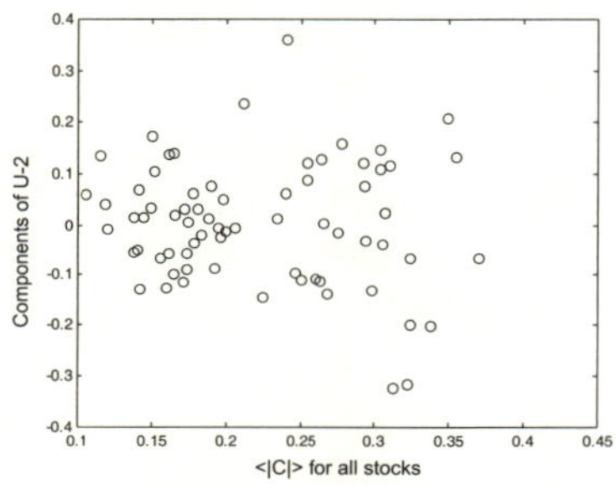

Fig. 11. Plot of the components of eigenvector U^2 associated with an eigenvalue from the bulk of RMT, λ_2. The variation shows no significant dependence between the two. The picture is quite the same for successive time periods considered.

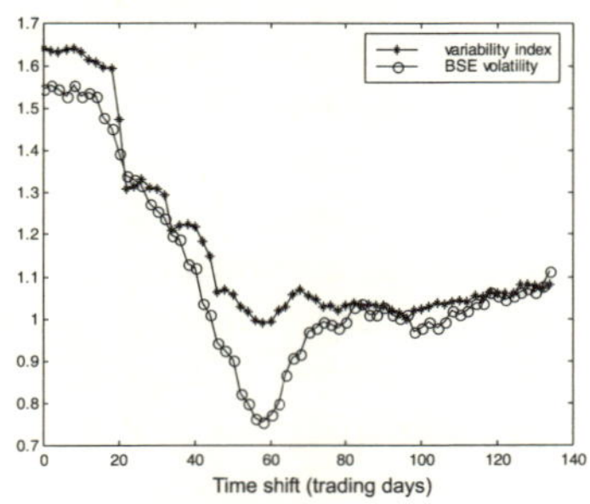

Fig. 12. Temporal evolution of the variability index, V and the volatility of BSE index is shown upon suitable scaling. The results are obtained from correlation matrix C constructed from daily returns of 70 stocks for 68 progressing time windows of 85 days each. The time was shifted in steps of 2 days each time and the time shift from the starting point is plotted on the horizontal axis.

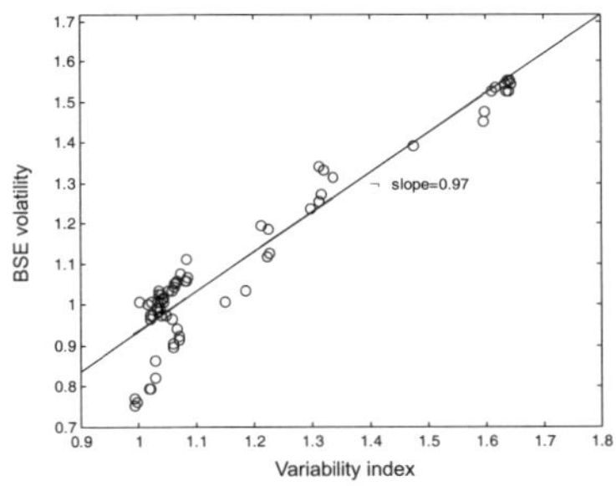

Fig. 13. The variability index, V with the volatility of BSE index approximates a linear fit with slope= 0.97 ± 0.04

4 Conclusions

We conclude with the following:
(i). Deviations from RMT bounds are more pronounced in volatile time periods as compared to the not so volatile ones in the context of Bombay Stock Exchange.
(ii). The largest eigenvalue, is seen to be highly sensitive to the trends of market activity. $\lambda_{max} * BSE\ volatility >= 0.94$.
(iii). Strong dependence between the average of magnitude of elements of C and volatility, indicating highly synchronous movements of stocks in highly fluctuating times or vice versa. $< |C| * BSE\ volatility >= 0.94$.
(iv). Role of the 'last' eigenmode in quantifying the fluctuations. Our finding on probability density patterns of U^{70} may suggest that "ensemble-like" behavior is more prominent in volatile situations than non-volatile ones.
(v). A strong anti-correlation between IPR and volatility $(= -0.63)$ confirms the existence of a positive association between the number of significant participants in U^{70} with the volatility, see reference [15].
(vi). It is verified that the eigenvector U^{70} indicates the extent to which the stock movements are synchronized. Finally, the 'last' eigenstate of the cross correlation matrix is set up usefully to obtain a quantity that has statistically significant predictive power for the variability of the market at any time, the *Variabilty index*, V. The evolution of V and BSE index volatility, have identical trends, $< V * BSE\ volatility >= 0.95$.

References

1. Bachelier L (1900) 'Theorie de la speculation' [Ph.D. thesis in mathematics], Annales Scientifiques de l'Ecole Normale Superieure III-17, pp.21-86; Kulkarni S (1978) Sankhya: The Indian Journal of Statistics V40 Series D

2. Mantegna RN, Stanley HE (2000) An Introduction to Econophysics, Cambridge University Press
3. Mehta M (1991) Random Matrices, Academic Press
4. Laloux L, Cizeau P, Bouchaud J-P, Potters M (1999) Physical Review Letters 83:1467; (1999) Risk 12(3):69; Plerou V, Gopikrishnan P, Rosenow B, Amaral LAN, Stanley HE (1999) Phys. Rev. Lett. 83:1471; Potters M, Bouchaud J-P, Laloux L, arXiv:physics/0507111.
5. Liu Y, Gopikrishnan P, Cizeau P, Meyer M, Peng CK, Stanley HE (1999) Physical Review E 60(2)
6. Lillo F, Mantegna RN (2000) Physical Review E 62:6126-6134 Lillo F, Bonanno G, Mantegna RN (2002) Ed. by H. Takayasu, Springer-Verlag Tokyo; Micciche S, Bonanno G, Lillo F, Mantegna RN (2002) Physica A 314:756-761; Lillo F, Mantegna RN (2005) Physical Review E 72:016219
7. Drozdz S, Grummer F, Gorski AZ, Ruf F, Speth J (2000) Physica A 287:440; Drozdz S, Grummer F, Ruf F, Speth J (2001) Physica A 294:226; Empirical Science of Financial Fluctuations, H. Takayasu (ed.), Springer-Verlag Tokio 2002, p.41.
8. Gopikrishnan P, Rosenow B, Plerou V, Stanley HE (2001) Physical Review E 64:035106
9. Plerou V, Gopikrishnan P, Rosenow B, Amaral LAN, Guhr T, Stanley HE (2002) Physical Review E 65:066126
10. Burda Z, Jurkiewiez J, Nowak MA, Papp G, Zahed I (2001) cond-mat/0103108; Burda Z, Goerlich A, Jarosz A, Jurkiewicz J (2003) cond-mat/0305627
11. Wilcox D, Gebbie T (2004) arXiv:cond-mat/0402389; arXiv:cond-mat/0404416
12. Gençay R, Dacorogna M, Muller U, Pictet O, Olsen R (2001) An Introduction to High Frequency Finance, Academic Press Inc. (London)
13. Shalen C (1999) International Stock Index Spread Opportunities in Volatile Markets, The CBOT Dow Jones Industrial Average Futures and Future Options, 02613
14. Mounfield C, Ormerod P (2001) Market Correlation and Market Volatility in US Blue Chip Stocks, Crowell Prize Submission, 2001
15. Kulkarni V, Deo N (2005) arXiv:Physics/0512169.

Why do Hurst Exponents of Traded Value Increase as the Logarithm of Company Size?

Zoltán Eisler[1] and János Kertész[1,2]

[1] Department of Theoretical Physics, Budapest University of Technology and Economics, Budafoki út 8, H-1111 Budapest, Hungary
[2] Laboratory of Computational Engineering, Helsinki University of Technology, P.O.Box 9203, FIN-02015 HUT, Finland

Summary. The common assumption of universal behavior in stock market data can sometimes lead to false conclusions. In statistical physics, the Hurst exponents characterizing long-range correlations are often closely related to universal exponents. We show, that in the case of time series of the traded value, these Hurst exponents increase logarithmically with company size, and thus are non-universal. Moreover, the average transaction size shows scaling with the mean transaction frequency for large enough companies. We present a phenomenological scaling framework that properly accounts for such dependencies.

Key words: econophysics; stock market; fluctuation phenomena

1 Introduction

The last decades have seen a lot of contribution by physicists to various other subjects. The applied methods are often rooted in modern statistical physics, and in particular scaling and universality. Examples range from biology [1] to finance [2–5]. Successes achieved in the latter area have lead to the formation of a whole new field, commonly called econophysics. But despite the large number of studies and the undeniable progress that has been made, one worry still remains: There is very little theoretical ground to assume that physical concepts are actually appropriate to describe, e.g., stock market fluctuations [6]. Critically speaking, it is equally justified to consider the often used power laws as only a way to fit data. Instead of universality, what one actually observes in econophysics, can also be seen as just a robustness of qualitative features, which is a much weaker property.

In this paper we revisit a previously introduced framework for financial fluctuations [7], that can be used to explicitly show the absence of universal behavior in trading activity. The paper is organized as follows. Section 2 introduces notations and the dataset that will be used. Section 3 shows, that many key features of the traded value of stocks depend on the size of the company whose stock is considered. We find that as we go to larger and larger companies:

(i) the average transaction size increases,
(ii) the Hurst exponent of traded value/min grows as the logarithm of the mean of the same quantity,
(iii) fluctuations of the trading activity grow as a non-trivial, time scale dependent power of mean traded value.

Section 4 integrates these findings into a consistent, common scaling framework, and points out the connection between the three observations.

2 Notations and data

First, let us introduce a few notations that will be used throughout the paper. For a time window size Δt, one can write the total traded value of stock i during the interval $[t, t+\Delta t]$ as

$$f_i^{\Delta t}(t) = \sum_{n, t_i(n) \in [t, t+\Delta t]} V_i(n), \tag{1}$$

where $t_i(n)$ is the time of the n-th transaction of stock i. The number of elements in the sums, i.e., the number of trades in the time window, we will denote as $N_{\Delta t}(t)$. The so called tick-by-tick data are denoted by $V_i(n)$, which is the value exchanged in trade n. This is the product of the transaction price $p_i(n)$ and the traded volume $\tilde{V}_i(n)$:

$$V_i(n) = p_i(n)\tilde{V}_i(n). \tag{2}$$

Note that the use of V instead of $\tilde{V}$ automatically eliminates any anomalies caused by stock splits or dividends.

The data we analyze is from a TAQ database [8], containing all transactions of the New York Stock Exchange (NYSE) for the years 1993 – 2003. The samples were always restricted to those stocks that were traded every month during the period of that specific calculation. We detrended the data by the well-known U-shaped daily pattern of traded volumes, similarly to Ref. [9].

Finally, $\langle\cdot\rangle$ always denotes time average, and $\log(\cdot)$ means 10-base logarithm throughout the paper.

3 Size dependent properties of trading activity

In the econophysics literature, it is common practice to assume a form of universal behavior in stock market dynamics. The trading of different stocks, on different markets and for various time periods is assumed to follow the same laws, and this is – at least qualitatively – indeed found in the case of many stylized facts [4, 5]. However, recent studies [7, 10] have pointed out, that this is not completely general. In this section, we present two properties of trading, that appear robust between markets and time periods, and which are related to a distinct company size dependence.

Company size is usually measured by the capitalization, but trading frequency $\langle N_{\Delta t}\rangle$ (measured in trades/min), or the average traded value $\langle f_{\Delta t}\rangle$ (measured in USD/min) are also adequate measures of the importance of a company: Very small companies are traded infrequently, while large ones very often, and, naturally, traded value has a corresponding behavior. In fact, one finds, that $\langle N_{\Delta t}\rangle$ and $\langle f_{\Delta t}\rangle$ are non-trivial, monotonic functions of capitalization [10, 11].

3.1 Dependence of the average trade size on trading frequency

Let us first construct a very simple measurement: calculate the average number of trades per minute ($\langle N_i\rangle$) and the mean value exchanged per trade ($\langle V_i\rangle$). One can plot these two quantities versus each other for all stocks (see Fig. 1), to find a remarkably robust behavior. For all the periods 1994 – 1995, 1998 – 1999, and 2000, the data lack a clear tendency where trading frequency is low ($\langle N_i\rangle < 10^{-2}$ trades/min). Then, as we go to more frequently traded companies, an approximate power law emerges:

$$\langle V_i\rangle \propto \langle N_i\rangle^{\beta} . \tag{3}$$

The measured exponents are around $\beta \approx 0.5$, systematically greater than the value $\beta \approx 0.2$ found for NASDAQ (see also Refs. [7, 10]), and smaller than $\beta \approx 1$ for London's FTSE-100 [11].

In some sense trades appear to "stick together": Once a stock is traded more and more intensively, traders seem to prefer to increase their size as the frequency cannot be increased beyond limits.

3.2 Size-dependent correlations

The correlation properties of stock market time series have been studied extensively [4, 5, 12, 13]. However, with very few exceptions [14], such studies were limited to the stocks of large companies. Those, in general, were found to display universal patterns of behavior.

In this section we focus on the correlations of the traded value f. Recently it was pointed out by two independent studies [10, 15] that both this f and trading volumes have finite variance, in contrast to early findings [12]. Thus, it is meaningful to define a Hurst exponent $H(i)$ [16, 17] for f as

$$\sigma(i, \Delta t) = \left\langle \left(f_i^{\Delta t}(t) - \langle f_i^{\Delta t}(t)\rangle .\right)^2 \right\rangle \propto \Delta t^{H(i)}, \tag{4}$$

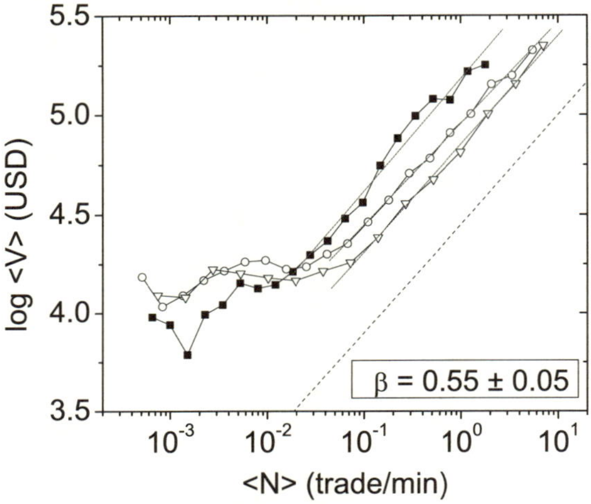

Fig. 1. The dependence of the mean value per trade $\langle V_i \rangle$ on the average number of trades/min $\langle N_i \rangle$. Calculations were done for the periods 1994 – 1995 (□), 1998 – 1999 (O), and for the year 2000 (▽, see Ref. [10]). For the smallest stocks the data lack a clear tendency. However, larger stocks show scaling between the two quantities, according to (3). The slopes are around $\beta = 0.55 \pm 0.05$, regardless of time period. *Note*: Groups of stocks were binned, and log $\langle V_i \rangle$ was averaged for better visibility.

The signal is correlated for $H > 0.5$, and uncorrelated for $H = 0.5$. Significant anticorrelated behavior ($H < 0.5$) does not usually occur in this context.

One finds, that the Hurst exponent does not exist in a strict sense: all stocks show a crossover [10] between two types of behavior around the time scale of 1 day. This threshold depends on the market and the time period under study, but keeping those constant, it does not depend on the actual stock in any systematic way.

We did the calculations for two time periods, 1994 – 1995 and 1998 – 1999. Under a certain size of time windows, which is $\Delta t < 20$ min for 1994 – 1995 and $\Delta t < 6$ min for 1998 – 1999, the trading activity is uncorrelated for all stocks. However, when one chooses $\Delta t > 300$ min, the picture changes completely. There, small $\langle f \rangle$ stocks again display only very weak correlations, but larger ones up to $H \approx 0.9$. Moreover, there is a clear logarithmic trend in the data:

$$H(i) = H^* + \gamma_t \log \langle f_i \rangle \,, \tag{5}$$

with $\gamma_t(\Delta t > 300 \text{ min}) = 0.05 \pm 0.01$ for 1994 – 1995 and $\gamma_t(\Delta t > 300 \text{ min}) = 0.07 \pm 0.01$ for 1998 – 1999. As a reference, we also checked that $H_{\text{shuff}}(i) = 0.5$ for the shuffled time series. All results are shown in Fig. 2.

The most interesting point is that the crossover is not from *uncorrelated* to *correlated*, but from *homogeneous* to *inhomogeneous* behavior. For short times, all stocks show $H(i) \approx H_1$, i.e., $\gamma_t = 0$. For long times, $H(i)$ changes

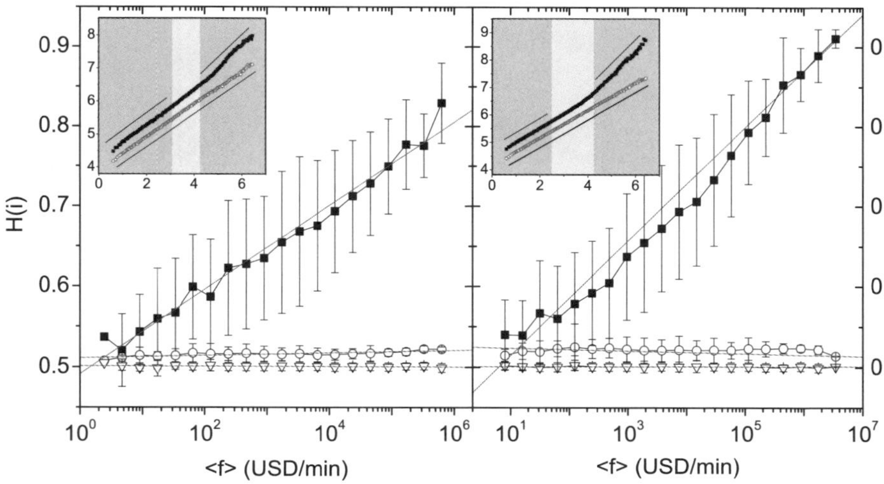

Fig. 2. Value of the Hurst exponents $H(i)$ for the time periods 1994–1995 (left) and 2000 – 2002 (right). For short time windows (O, $\Delta t < 20$ min for 1994 – 1995, and $\Delta t < 6$ min for 1998 – 1999), all signals are nearly uncorrelated, $H(i) \approx 0.51 - 0.52$. The fitted slope is $\gamma_t = 0.00 \pm 0.01$. For larger time windows (■. $\Delta t > 300$ min), the strength of correlations depends logarithmically on the mean trading activity of the stock, $\gamma_t = 0.05 \pm 0.01$ for 1994 – 1995, and $\gamma_t = 0.05 \pm 0.01$ for 1998 – 1999. Shuffled data (▽) display no correlations, thus $H_{\text{shuff}}(i) = 0.5$, which also implies $\gamma_t = 0$. *Note*: Groups of stocks were binned, and their logarithm was averaged. The error bars show standard deviations in the bins. *Insets:* The $\log \sigma$-$\log \Delta t$ scaling plots for Wal-Mart (WMT, ■). The darker shaded intervals have well-defined Hurst exponents, the crossover is indicated with a lighter background. The slopes corresponding to Hurst exponents are 0.52 and 0.73 for 1994 – 1995, and 0.52 and 0.89 for 1998 – 1999. The slope for shuffled data is 0.5. Shuffled points (O) were shifted vertically for better visibility.

with $\langle f_i \rangle$ and $\gamma_t > 0$. This can also be understood as a dependence on company size, as $\langle f \rangle$ is roughly proportional to capitalization [10].

3.3 Fluctuation scaling

The technique of fluctuation scaling is very similar to the above, and it was recently applied to stock market data (see, e.g., Refs. [7, 9]). It is based on a phenomenological scaling law that connects the standard deviation σ_i and the average $\langle f_i \rangle$ of the trading activity for all stocks:

$$\sigma(i, \Delta t) \propto \langle f_i \rangle^{\alpha(\Delta t)}, \tag{6}$$

where the scaling variable is $\langle f_i \rangle$ (or i), and Δt is kept constant. That is, the standard deviation of a quantity scales with the mean of the same quantity.

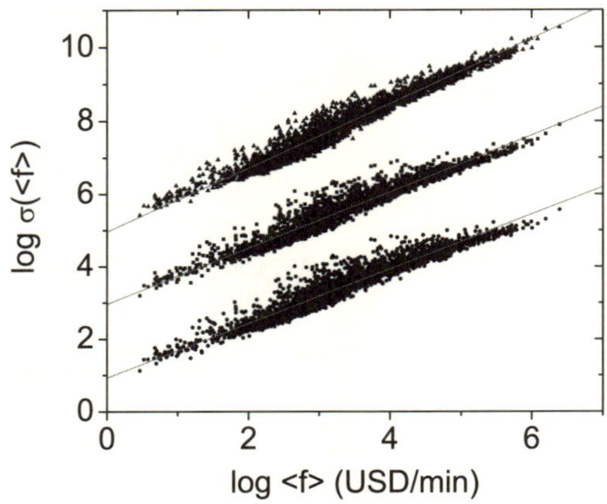

Fig. 3. Examples of $\sigma(\langle f \rangle)$ scaling plots for NYSE, period 1998 – 1999. The window sizes from bottom to top: Δt = 10 sec, 0.5 day, 2 weeks. The slopes are $\alpha(\Delta t) = 0.75, 0.78, 0.88$, respectively. Points were shifted vertically for better visibility.

$\sigma(i, \Delta t)$ is the same as used in the definition of the Hurst exponent (4), where i was constant and Δt was varied.

The presence of scaling (6) is not at all a trivial fact. Nevertheless, one finds that it holds quite generally, for any Δt. Here, we confirm this for the periods 1994 – 1995 and 1998 – 1999, examples of scaling plots are shown in Fig. 3.

A systematic investigation finds, that α strongly depends on the Δt size of the time windows. Fig. 4 shows, that when Δt is at most a few minutes, $\alpha(\Delta t)$ is constant, the values are 0.74 and 0.70 for 1994 – 1995 and 1998 – 1999, respectively. Then, after an intermediate regime, for window sizes $\Delta t > 300$ min, there is a logarithmic trend:

$$\alpha(\Delta t) = \alpha^* + \gamma_f \log \Delta t, \tag{7}$$

with slopes $\gamma_f = 0.05 \pm 0.01$ for 1994 – 1995, and $\gamma_f = 0.07 \pm 0.01$ for 1998 – 1999.

4 Scaling theory

The stock market is only one of many examples for fluctuation scaling, which is in fact a very general and robust phenomenon, that was observed in many complex systems, ranging from highway traffic to fluctations in the visitations of web pages [9, 18, 19]. Here, the elements of the system are the web pages or highways, and $f_i(t)$ is not trading activity, but still some a form "activity" (number of visitation to the page, volume of car traffic through the road).

A previous study [18] found, that in temporally uncorrelated systems, two universality classes exist. Under strong external driving, all systems show

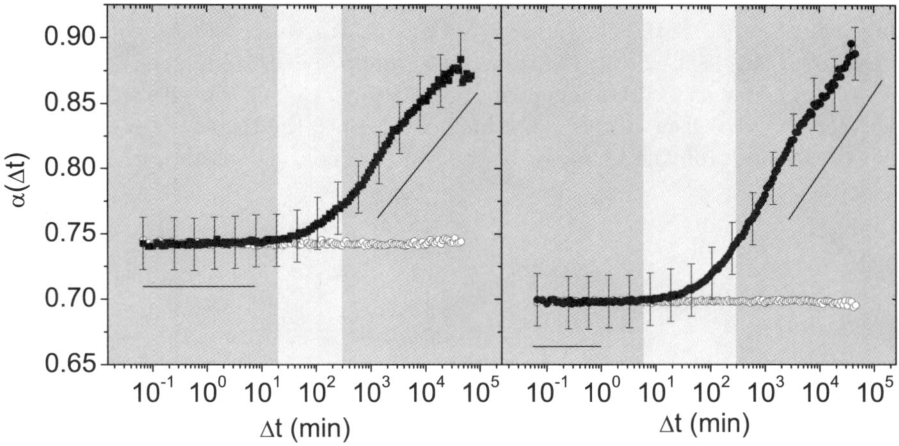

Fig. 4. The dependence of the scaling exponent α on the window size Δt. The darker shaded intervals have well-defined Hurst exponents and values of γ_t, the crossover is indicated with a lighter background. 1994 – 1995 (left): without shuffling (■) the slopes of the linear regimes are $\gamma_f(\Delta t < 20 \text{ min}) = 0.00 \pm 0.01$ and $\gamma_f(\Delta t > 300 \text{ min}) = 0.05 \pm 0.01$. For shuffled data (O) the exponent is independent of window size, $\alpha(\Delta t) = 0.74 \pm 0.02$. 1998–1999 (right): without shuffling (■) the slopes of the linear regimes are $\gamma_f(\Delta t < 6 \text{ min}) = 0.00 \pm 0.01$ and $\gamma_f(\Delta t > 300 \text{ min}) = 0.07 \pm 0.01$. For shuffled data (O) the exponent is independent of window size, $\alpha(\Delta t) = 0.70 \pm 0.02$.

$\alpha = 1$. Systems with a robust internal dynamics, consisting of i.i.d. events, display $\alpha = 1/2$.

When the size of the events is not identically distributed throughout the system, that can lead to the breaking of universality, and intermediate values of α. This is what happens in the case of stock markets. When Δt is small (seconds to a few minutes), the transactions can be assumed to arrive independently, but their size is inhomogeneous, as pointed out in Sec. 3.1. In the complete absence of correlations, there would a clear relationship between the exponents α and β [19]:

$$\alpha = \frac{1}{2}\left(1 + \frac{\beta}{\beta + 1}\right). \tag{8}$$

Substituting $\beta = 0.55$ yields $\alpha(\Delta t \to 0) = 0.68 \pm 0.01$, which agrees within the error bars with the result for 1998 – 1999, but it is somewhat smaller than the actual value for 1994 – 1995. Also note that β only exists for large enough stocks, whereas α and fluctuation scaling applies to all stocks. We believe, that the discrepancies are due to the fact, that the picture presented in Ref. [19] is an overly simplified model for the stock market. Nevertheless, it is remarkable, that the breaking of universality and the rough value of the exponent is predicted properly.

Let us now turn to the Δt dependence of α. First of all, let us notice, that for both periods, there is a change from a constant value to a logarithmic increase, and this is at exactly the same Δt's, where Sec. 3.2 found the crossovers from homogeneous to inhomogeneous correlations. In fact, the correspondence between the two observations is not incidental. Both the Hurst exponents $H(i)$ and $\alpha(\Delta t)$ describe the behavior of the same standard deviation $\sigma(i, \Delta t)$:

$$\sigma(i, \Delta t) = \left\langle \left(f_i^{\Delta t}(t) - \langle f_i^{\Delta t}(t)\rangle\right)^2 \right\rangle \propto \Delta t^{H(i)},$$

and

$$\sigma(i, \Delta t) \propto \langle f_i \rangle^{\alpha(\Delta t)}.$$

A simple calculation [7] can show that the only possible way for these two scaling laws to coexist, is when

$$\alpha(\Delta t) = \alpha^* + \gamma \log \Delta t \tag{9}$$

$$H(i) = H^* + \gamma \log \langle f_i \rangle, \tag{10}$$

where a key point is that the two slopes γ are the same. In short, for the two previously introduced constants $\gamma_t = \gamma_f = \gamma$.

Again, this is in harmony with the actual observations. Due to the previously mentioned crossover in correlations, one has to distinguish three regimes in Δt.

1. For small Δt, all stocks display the same, nearly uncorrelated trading behavior, i.e., $\gamma = 0$. Accordingly, $\alpha(\Delta t)$ is constant, regardless of window size.
2. For an intermediate range of Δt's, we are in the crossover regime. H does not exist for any stock. α still does exist, but – as expected – its time window dependence does not follow a logarithmic trend.
3. For large Δt, the Hurst exponent increases logarithmically with the mean traded value $\langle f \rangle$, and so does α with Δt. The slopes agree very well ($\gamma_t = \gamma_f$) for both time periods.

As noted before, the equality $\gamma_t = \gamma_f$ can be calculated fairly easily, but one can look at this result in a different way. Both fluctuation scaling and the Hurst exponent (or equivalently, power law autocorrelations) are present in a very wide range of complex systems. But we have just seen that this is only possible in two of ways: the correlations must either be homogeneous throughout the system ($H(i) = H$, $\gamma = 0$), or they must have a logarithmic dependence on mean activity. Consequently, when one for example looks at the results of Sec. 3.2, they are not surprising at all. The coexistence of our two scaling laws is so restrictive, that if the strength of correlations depends on company size, and thus on $\langle f \rangle$, the realized logarithmic dependence is the only possible scenario.

5 Conclusions

In the above, we presented some recent findings concerning the fluctuations of stock market trading activity. As the central point of the paper, we discussed the application of fluctuation scaling. We gave evidence, that the logarithmic increase of the Hurst exponent of traded value with the mean traded value comes as a very natural consequence of fluctuation scaling. The behavior of the companies depends on a continuous parameter: the average traded value, or equivalently, company size.

This is a clear sign of non-universality, and thus contrary to a naive expectation from statistical physics. For example, in the case of surface growth [16], the Hurst exponent of surface height fluctuations is universal to a certain type of growth dynamics. In contrast, on the market the "dynamics" (i.e., trading rules) are the same for all stocks, but the resulting exponents still vary. While we believe that it is possible, that at least some properties of trading are universal, but we wish to point out that not *all* of them are.

Our results imply that one must take great care when applying concepts like scaling and universality to financial markets. The present theoretical models of trading should be extended to account for the capitalization dependence of the characteristic quantities, which is a great challenge for future research.

Acknowledgments

The authors would like to express their gratitude to Bikas K. Chakrabarti, Arnab Chatterjee and all organizers of the International Workshop on the Econophysics of Stock Markets and Minority Games for their infinite hospitality. They are also indebted to György Andor for his support with financial data. JK is member of the Center for Applied Mathematics and Computational Physics, BME. Support by OTKA T049238 is acknowledged.

References

1. T. Vicsek, editor. *Fluctuations and Scaling in Biology.* Oxford University Press, USA, 2001.
2. P.W. Anderson, editor. *The Economy As an Evolving Complex System (Santa Fe Institute Studies in the Sciences of Complexity Proceedings)*, 1988.
3. J. Kertész and I. Kondor, editors. *Econophysics: An Emergent Science, http://newton.phy.bme.hu/~kullmann/Egyetem/konyv.html.* 1997.
4. J.-P. Bouchaud and M. Potters. *Theory of Financial Risk.* Cambridge University Press, Cambridge, 2000.
5. R.N. Mantegna and H.E. Stanley. *Introduction to Econophysics: Correlations and Complexity in Finance.* Cambridge University Press, 1999.
6. M. Gallegatti, S. Keen, T. Lux, and P. Ormerod. Worrying trends in econophysics. *http://www.unifr.ch/econophysics, doc/0601001; to appear in Physica A, Proceedings of the World Econophysics Colloquium, Canberra, 2005.*

7. Z. Eisler and J. Kertész. Scaling theory of temporal correlations and size dependent fluctuations in the traded value of stocks. *arXiv:physics/0510058*, 2005. to appear in Phys. Rev. E.
8. Trades and Quotes Database for 1993-2003, New York Stock Exchange, New York.
9. Z. Eisler, J. Kertész, S.-H. Yook, and A.-L. Barabási. Multiscaling and non-universality in fluctuations of driven complex systems. *Europhys. Lett.*, 69:664–670, 2005.
10. Z. Eisler and J. Kertész. Size matters: some stylized facts of the market revisited. *arXiv:physics/0508156*, 2005.
11. G. Zumbach. How trading activity scales with company size in the FTSE 100. *Quantitative Finance*, 4:441–456, 2004.
12. P. Gopikrishnan, V. Plerou, X. Gabaix, and H.E. Stanley. Statistical properties of share volume traded in financial markets. *Phys. Rev. E*, 62:4493–4496, 2000.
13. R. Cont. Empirical properties of asset returns: stylized facts and statistical issues. *Quantitative Finance*, 1:223–236, 2001.
14. G. Bonanno, F. Lillo, and R.N. Mantegna. Dynamics of the number of trades of financial securities. *Physica A*, 280:136–141, 2000.
15. S.M.D. Queirós. On the distribution of high-frequency stock market traded volume: a dynamical scenario. *Europhys. Lett.*, 71:339–345, 2005.
16. T. Vicsek. *Fractal Growth Phenomena.* World Scientific Publishing, 1992.
17. J.W. Kantelhardt, S.A. Zschiegner, E. Koscielny-Bunde, S. Havlin, A. Bunde, and H.E. Stanley. *Physica A*, 316:87–114, 2002.
18. M.A. de Menezes and A.-L. Barabási. Fluctuations in network dynamics. *Phys. Rev. Lett.*, 92:28701, 2004.
19. Z. Eisler and J. Kertész. Random walks on complex networks with inhomogeneous impact. *Phys. Rev. E*, 71:057104, 2005.

Statistical Distribution of Stock Returns Runs

Honggang Li[1] and Yan Gao[2]

[1] Department of Systems Science, School of Management, Beijing Normal University, Beijing 100875, China. hli@bnu.edu.cn
[2] Department of Systems Science, School of Management, Beijing Normal University, Beijing 100875, China. haoy@163.com

Summary. In this paper, we focus on the statistical features of runs which is defined as a sequence of consecutive gain/loss (rise/fall) stock returns. By studying daily data of the Dow Jones industrial average (DJIA), we get the following points: firstly, the distribution of the length and magnitude of stock return runs both follow an exponential law; and secondly, the positive runs and negative runs show a significant asymmetry in frequency distribution. We expect that the two properties may be new members in the family of stylized facts about stock returns.

1 Introduction

Over the past decade, more and more empirical studies on financial time series have presented some stylized facts which are common properties across a wide range of markets, instruments and time periods [1], and recently, studies aimed to find new stylized facts still continue. For instance, exponential distribution of financial returns at mesoscopic time lags has been found by Silva, Prange and Yakovenko, and they recognize the exponential distribution as another stylized fact in the set of analytical tools for financial data analysis [2].

Generally, studies on stock returns have their eyes on the value of individual return with certain time horizon such as hourly, daily and weekly. But we think the sign of returns, namely just the rise or fall property of stock price, is also a good indicator for stock price movement. Specially, studying a sequence of signs of returns may bring us some nontrivial information on market evolution. Furthermore, instead of the individual return, taking the accumulative return of a specific sequence as a target to explore may be a good alternative. In fact, this paper just wants to give a try of empirical work in this way.

A sequence of consecutive positive or negative stock returns has a term as "runs". Furthermore, here we define positive runs and negative runs, namely that positive runs is a sequence of positive stock returns whereas negative runs is a sequence of negative stock returns. For example, a particular sequence of 10 returns(1%, −2%,−3%,4%,5%,6%,−7%,8%,−9%,−10%),using the sign of their values, may be represented by "+ − − + + + − + − −", containing 3 positive runs of +s and 3 negative runs of -s, thus 6 runs in total. The length of a runs is the number of observations included in the runs, and the magnitude of a runs is defined as the accumulative return of a runs. Considering the mentioned example, the length of 3 positive runs are 1, 3, 1 respectively, and magnitude are 1%,15.75%,8% respectively; the length of 3 negative runs are 2, 1, 2, respectively, and magnitude are −4.94%,−7%,−18.1% respectively.

Based on these definitions, we explored the DJIA return data and found the distribution of the length and magnitude of runs in stock returns both follow an exponential law. Furthermore, we compared the frequency distributions of positive runs and negative runs with different length and found they show a significant difference, this may imply another evidence of gain/loss asymmetry in stock returns [1].

2 Statistical analysis on length of runs

The raw daily price index data on DJIA is taken from Yahoo financial database and the return is calculated in a usual way as

$$r_t = (p_t - p_{t-1})/p_{t-1}$$

where p_t is the close price index at time t. We get 19397 returns from 19398 transaction dates data from October 1 of 1928 to December 30 of 2005.

2.1 Runs statistics and runs test

Using the selected sample returns, positive runs and negative runs we once defined were counted. In the 19397 returns, there are 10154 positive returns and 9243 negative returns, and both the number of positive runs and negative runs are 4597. Length of the longest positive runs is 13, corresponding to the date period from January 2 to January 20 of 1987, and length of the longest negative runs is 12, corresponding to one period from July 29 to August 13 of 1941 and another from January 9 to January 24 of 1968. Table 1 gives frequency statistics of runs by its length.

Table 1. Frequencies of positive and negative runs with different length

length	1	2	3	4	5	6	7	8	9	10	11	12	13
positive runs	1976	1199	684	353	184	112	37	25	15	6	3	2	1
negative runs	2215	1193	612	300	156	67	32	12	3	2	3	2	0

To test randomness of the return time series, here the runs test [3] was done. The SPSS tells the z value is -6.967 and the Asymp. Sig. (2-tailed) is

0.000, which both mean the return series exists autocorrelation at 1% significance level.

2.2 Statistical distributions of runs

The relationship between the frequency, N, of the observations and the length, L, in days of a runs is postulated to be of the form of an exponential law

$$\mathbf{N} = \mathbf{ae}^{-\mathbf{bL}} \tag{1}$$

After taking a ln-linear transformation, the parameters in (1) could be estimated by a linear regression which yields a=4416.046 and b= -0.648 for positive runs; while a=4491.761 and b= -0.708 for negative runs. R^2 values of the above two regression functions are 0.997 and 0.970 respectively which mean (1) describes the relationship between frequency and length of runs fairly well. Fig.1.a shows the result visually.

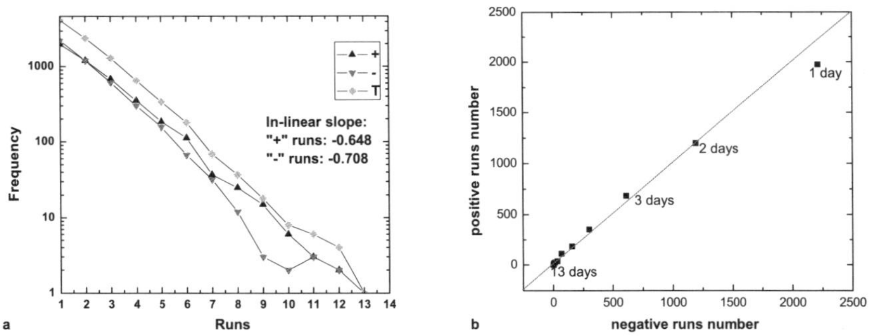

Fig. 1. a. Exponential distribution of frequency on length of runs: with a fit slope -0.648 for positive runs and with a fit slope -0.708 for negative runs. The up-triangle points denote positive runs, down-triangle points denote negative runs and square points denote sum of positive runs and negative runs. b.Comparison of frequency distribution between positive runs and negative runs with different length.

Previous research has found gains and losses are asymmetric in stock market, namely that one observes large draw downs in stock index values but not equally large upward movements. Then, can we get some similar findings by runs analysis? Here we have a try in a direct way by comparing frequency distribution of positive runs, which corresponds to gains sequence, and negative runs, which corresponds to loss sequence. Fig.1.b plots the positive runs number and negative runs number with the same runs length. With a perfect symmetry distribution, all the points in the chart would lie on the 45° line. However, it is clear that the real frequency distribution suggests that there exists a significant asymmetry of positive runs and negative runs. This just implies an asymmetry on persistence of gains and losses in stock market. Specially, with length 1, there are more negative runs than positive runs; but with length more than 1, there are more positive runs than negative runs. This just means daily gains have a better persistence than daily losses for our sample.

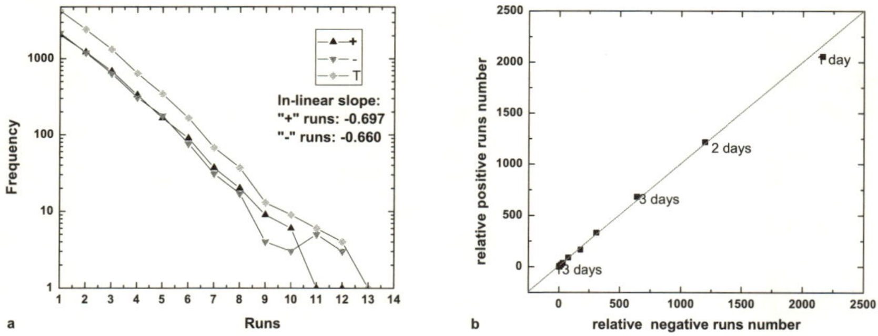

Fig. 2. a.Exponential distribution of frequency on length of runs: with a fit slope -0.697 for relative positive runs and with a fit slope -0.660 for relative negative runs. b.Comparison of frequency distribution between relative positive runs and relative negative runs with different length.

2.3 Statistical analysis considering the drift effect

As we all know, many historic returns time series posses a positive drift over long time scales. If that happens, one should hesitate to compare directly the positive and negative levels of return when we discuss rise and fall of stock price [4]. So, to remove such a drift effect in the analyzed time series, we can set the mean (0.02647%)of the sample returns as a cut point to define relative positive runs and relative negative runs. Similarly, relative positive runs is a sequence of stock returns which are successively above or equal to return mean whereas relative negative runs is a sequence of stock returns which are successively below return mean . Under this new definition, similar findings have been obtained. In the 19397 returns, there are 9839 relative positive returns and 9558 relative negative returns. Both the number of relative positive runs and the number of relative negative runs are 4621. Table 2 presents frequency statistics of relative runs by its length. The runs test also reports the return series exists autocorrelation at 1% significance level.

We make the same postulation as (1). Similarly, after taking a ln-linear transformation, we can also estimate the parameters in (1) by a linear regression which yields a=4979.078 and b= -0.697 for positive runs; while

Table 2. Frequencies of relative positive and negative runs with different length

length	1	2	3	4	5	6	7	8	9	10	11	12	13
positive runs	2056	1218	683	333	166	90	37	20	9	6	1	1	1
negative runs	2161	1196	639	308	178	76	31	17	4	3	5	3	0

a=3967.931 and b= -0.660 for negative runs. R^2 values of the above two regression functions are 0.987 and 0.964 respectively which mean (1) also describes the relationship between frequency and length of relative runs fairly well. Fig.2.a shows the result visually. Fig.2.b plots the relative positive runs number and relative negative runs number with the same runs length. It also shows that the real frequency distribution suggests a significant asymmetry of positive runs and negative runs and this means daily gains have a better persistence than daily losses.

2.4 Runs analysis based on one-day returns

If we calculate DJIA returns defined as one-day returns as

$$r_t = (p_2 - p_1)/p_1$$

where p_1 is the open price index and p_2 is the close price index. We can get one-day returns of 19398 transaction dates data during the above sample period. Then, using new data, repeating the work we have done can also get similar results. Fig.3.a and Fig.3.b reproduce the results.

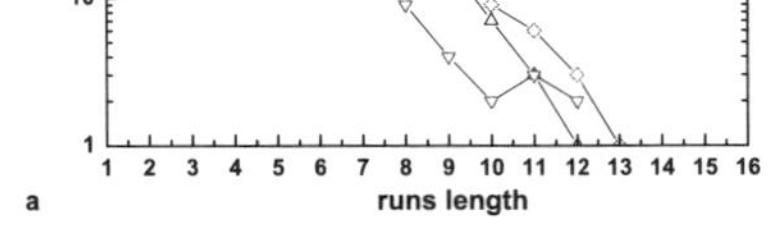

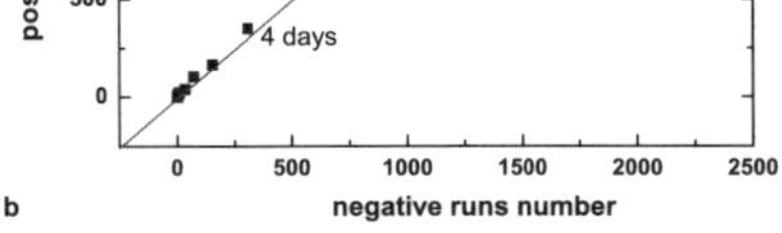

Fig. 3. a.Exponential distribution of frequency on length of runs: with a fit slope -0.662 for positive runs and with a fit slope -0.707 for negative runs. b.Comparison of frequency distribution between positive runs and negative runs with different length.

3 Statistical analysis on magnitude of runs

As one of stylized facts on stock market, the distribution of returns has a fat tail and the tail may follow a power law [1]. Fig.4.a shows the distribution of absolute value of positive returns and negative returns respectively. To compare with this distribution of individual returns, we here explore the distribution of cumulative returns of a runs.

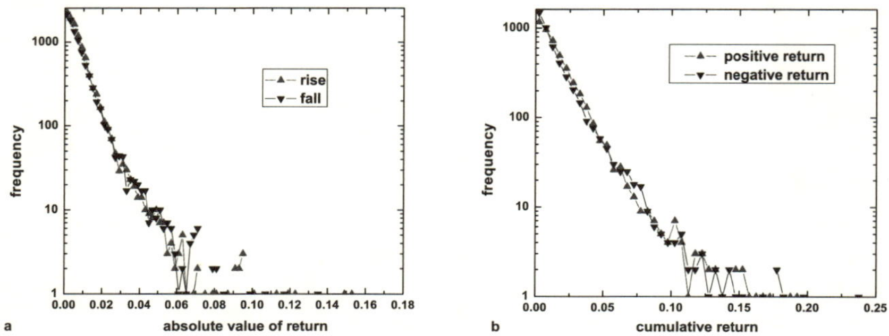

Fig. 4. a.Distributions of absolute value of positive returns and negative returns. b.Distribution of absolute value of cumulative return of a runs.

To see the statistical distribution property of magnitude, namely accumulative returns of stock return runs, we define accumulative return of a runs as $\prod_{\tau=t}^{t+lr}(1+r_\tau)-1$,where lr is length of the runs. For example, during a four-day positive runs from August 2 to August 8 of 1932, the return rose by a cumulative total of 27.37%,which corresponds to the biggest magnitude of runs in our sample returns.

By computing the accumulative returns of every runs, separating them to a series of positive groups and negative groups, counting the runs' numbers in each group, we can obtain frequency distribution of magnitude of runs. The result is shown in Table 3 and Fig.4.b, where it can be seen that the frequency distribution is significantly different from the distribution of daily returns shown in Fig.4.a, and it is nearly consistent with an exponential law. We can see the distribution also has a fat tail, but the tail does not fit power-law well. We test the result using the same means we did above, and get the parameters a=241.29 and b= -30.937 for positive runs; while a=316.081 and b= -29.005 for negative runs in (1).R^2 values of the above two regression functions are 0.792 and 0.802 respectively which mean (1) also describes the frequency distribution of magnitude of runs well.

Drawing the distribution of positive cumulative returns and negative cumulative returns together in one chart as Fig.5, a significant asymmetry in

Table 3. Frequencies of cumulative returns of positive runs and negative runs. The "c.return"denotes the median of cumulative return interval (in thousandth), "positive"and "negative"denote the number of positive and negative runs respectively.

c.return	5	10	15	20	25	30	34	40	45	50	55	60	65	70	75	80	85	90	95	100
positive	1164	956	724	495	356	246	184	130	85	55	49	26	28	17	13	9	9	7	5	4
negative	1473	1010	622	411	288	206	147	91	76	58	45	30	25	25	18	17	9	6	5	4

distribution of runs magnitude can also be seen. This asymmetry implies persistent fall with little cumulative loss is more frequent than persistent rise with little gain but persistent fall with big cumulative loss is less frequent than persistent rise with big gain for our sample.

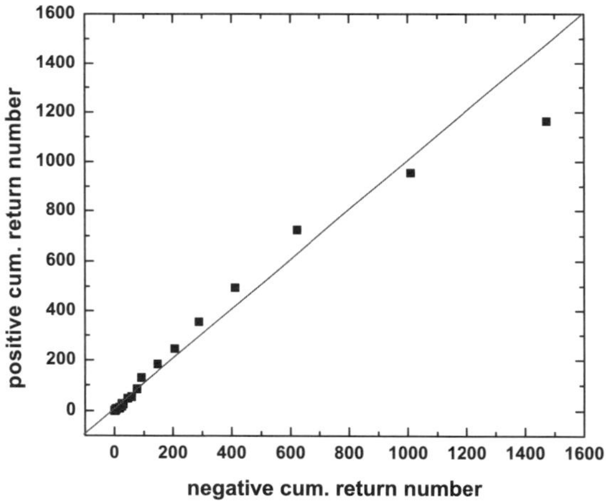

Fig. 5. Comparison between positive and negative runs with different absolute value of cumulative returns.

4 Conclusion

In this paper, we examined DJIA runs over a specific period. The aim is to explore some meaningful properties about the sequence of consecutive rise/fall of stock price by an empirical approach.

We found that the length of stock return runs follows an exponential distribution fairly well, and the magnitude of a runs also fits exponential law better than power law. Moreover, positive runs and negative runs show a significant asymmetry in frequency distribution.

The results of our study also give a new way to the research on persistence probability in market returns [5] and may add new members to the family of stylized facts.

Acknowledgments

We wish to thank Yougui Wang and Chunyan Lan for useful comments and discussions. None of the above are responsible for any errors in this paper. This work was supported by the National Science Foundation of China under Grant No. 70371073.

References

1. Cont R (2001) Quantitative Finance 1:223–236
2. Silva AC, Prange RE, Yakovenko VM (2004) Physica A 344:227–235
3. Campbell JY, Lo AW and Mackinlay AC (1997) The Econometrics of Financial Markets, Princeton University Press
4. Jensen MH, Johansen A, Simonsen I (2003) Physica A 324:338–343
5. Ren F, Zheng B, Lin H, Wen LY and Trimper S (2005) Physica A 350(2-4):439–450

Fluctuation Dynamics of Exchange Rates on Indian Financial Market

A. Sarkar and P. Barat

Variable Energy Cyclotron Centre, 1/AF Bidhan Nagar, Kolkata 700064, India.
apu@veccal.ernet.in, pbarat@veccal.ernet.in

Summary. Here we investigate the scaling behavior and the complexity of the average daily exchange rate returns of the Indian Rupee against four foreign currencies namely US Dollar, Euro, Great Britain Pound and Japanese Yen. Our analysis revealed that the average daily exchange rate return of the Indian Rupee against the US Dollar exhibits a persistent scaling behavior and follow Levy stable distribution. On the contrary the average daily exchange rate returns of the other three foreign currencies show randomness and follow Gaussian distribution. Moreover, it is seen that the complexity of the average daily exchange rate return of the Indian Rupee against US Dollar is less than the other three exchange rate returns.

1 Introduction

Foreign exchange market is the place where different currencies are traded, one for another. The exchange rate enables people in one country to translate the prices of foreign goods into units of their own currency. An appreciation of a nation's currency will make foreign goods cheaper. A depreciation of a nation's currency will make foreign goods more expensive. Under a flexible rate system, the exchange rate is determined by the supply and the demand. The Indian Rupee demand for the foreign exchange originates from Indian demand for foreign goods, services, and assets (real or financial). The supply of foreign exchange originates from sales of goods, services, and assets from Indians to foreigners.The foreign exchange market brings the quantity demanded and the quantity supplied into balance. As it does so, it brings the purchases by Indians from foreigners into equality with the sales of Indians to foreigners.Factors that cause a currency to appreciate are a slower growth rate relative to one's trading partners, a lower inflation than one's trading partners and an increase in domestic real interest rates (relative to rates abroad). On the other hand the factors responsible for the depreciation of a currency are a rapid growth of income (relative to trading partners) that stimulates imports relative to exports, a higher rate of inflation than one's trading partners and a reduction in domestic real interest rates (relative to rates abroad).

These different factors and the unknown nature of the interactions between them make the study of a foreign exchange market the most complicated and challenging one. In the recent years, new and sophisticated methods have been invented and developed in statistical and nonlinear physics to study the dynamical and structural properties of various complex systems. These methods have been successfully applied in the field of quantitative economy, which gave a chance to look at the economical and financial data from a new perspective [1]. The exchange rates between currencies are particularly interesting category of economic data to study as they dictate the economy of most countries. The time dependence of the exchange rates is usually complex in nature and hence, it is interesting to analyze using the newly developed statistical methods. In this paper we report the study of detailed scaling behavior of the average daily exchange rate returns of Indian Rupee (INR) versus four important foreign currencies in Indian economy, namely the US Dollar (USD), the EURO, the Great Britain Pound (GBP) and the Japanese YEN for the past few years. India, being the country with second largest population in the world, is an important business market for the multinational companies. Therefore the study of the average daily exchange rate returns of Indian Rupee with respect to the four foreign currencies is very significant and relevant from the economic point of view.

Scaling as a manifestation of underlying dynamics is familiar throughout physics. It has been instrumental in helping scientists gain deeper insights into problems ranging across the entire spectrum of science and technology. Scaling laws typically reflect underlying generic features and physical principles that are independent of detailed dynamics or characteristics of particular models. Scale invariance seems to be widespread in natural systems. Numerous examples of scale invariance properties can be found in the literature like earthquakes, clouds, networks etc. Scaling investigation in the financial data has been recently got much importance.

2 Data

We have studied the daily evolution of the currency exchange data of INR-USD, INR-EURO, INR-GBP and INR-YEN for the period of August 1998 to August 2004. The data were taken from the official website of the Reserve Bank of India [2].

The return Z(t) of the exchange rate time series X(t) is defined as:

$$Z(t) = \ln\left(\frac{X(t+1)}{X(t)}\right) \tag{1}$$

Fig. 1a and Fig. 1b show the variation of the average daily exchange rates of INR against USD and its return respectively.All the analyses reported in this paper were carried out on the return Z(t).

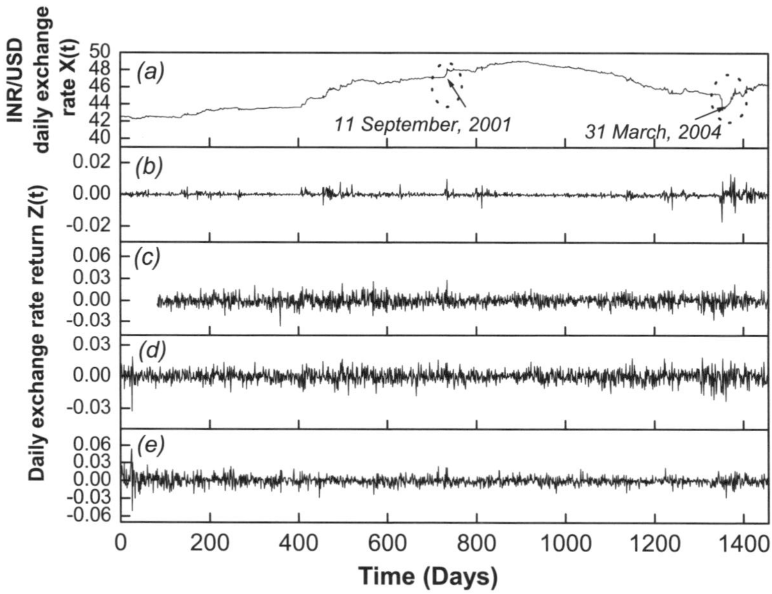

Fig. 1. (a)Variation of the average daily INR-USD exchange rate. Variation of the return of the average daily exchange rates of (b) INR-USD (c) INR-EURO (d) INR-GBP and (e) INR-YEN.

3 Method of analysis and results

3.1 Scaling analysis

We have used two newly developed methods of scaling analysis, namely (i) Finite Variance Scaling Method (FVSM) and (ii) Diffusion Entropy Analysis (DEA) [3] to detect the exact scaling behavior of the daily close price index of the BSE. These methods are based on the prescription that numbers in a time series R(t) are the fluctuations of a diffusion trajectory see Ref [3] for details. Therefore we shift our attention from the time series R(t) to probability density function (PDF) $p(x,t)$ of the corresponding diffusion process. Here x denotes the variable collecting the fluctuation and is referred to as the diffusion variable. The scaling property of $p(x,t)$ takes the form:

$$p(x,t) = \frac{1}{t^\delta} F\left(\frac{x}{t^\delta}\right) \qquad (2)$$

Finite variance scaling method

In the FVSM one examines the scaling properties of the second moment of the diffusion process generated by a time series. One version of FVSM is the standard deviation analysis (SDA), which is based on the eavaluation of the standard deviation *D(t)* of the variable x, and yeilds

$$D(t) = \sqrt{\langle x^2; t\rangle - \langle x; t\rangle^2} \propto t^\gamma \qquad (3)$$

The exponent γ is interpreted as the scaling exponent.

Diffusion entropy analysis

DEA introduced recently by Scafetta *et. al.* [4] focuses on the scaling exponent δ evaluated through the Shannon entropy *S(t)* of the diffusion generated by the fluctuations Z(t) of the time series using the scaling PDF. Here, the PDF of the diffusion process *p(x,t)* is evaluated by means of the subtrajectories $x_n(t) = \sum Z_{i+n}(t)$ with n=0,1,.... Using the scaling PDF (2) we arrive at the expression for *S(t)* as

$$S(t) = -A + \delta \ln(t) \qquad (4)$$

where A is a constant. Eq. (4) indiactes that in the case of a diffusion process with a scaling PDF, its entropy *S(t)* increases linearly with ln(t). Finally we compare γ and δ. For fractional Brownian motion the scaling exponent δ coincided with the γ. For random noise with finite variance, the PDF p(x,t) will converge to a Gaussian distribution with $\gamma = \delta = 0.5$. If $\gamma \neq \delta$ the scaling represents anomalous behavior.

The plots of SDA and DEA for the average daily exchange rate returns of the four foreign currencies are shown in Fig. 2 and Fig. 3 respectively.

The scaling exponents obtained from the plots of SDA and DEA are listed in Table 1. The values of and clearly reflect that the INR-USD exchange rate returns behave in a different manner with respect to the other three exchange rate returns. For INR-USD exchange rate returns the scaling exponents are found to be greater than 0.5 indicating a persistent scaling behavior. While the unequal values of and implies anomalous scaling. For the other three exchange rate returns, the values of and are almost equal to 0.5 within their statistical error limit, signifying absence of scaling in those cases.

The primary objectives of these analyses were to find the generic feature of these time series data, their long range correlation and their robustness to retain the scaling property. To verify the robustness of the observed scaling property of INR-USD exchange rate return data, we corrupted 2% of the

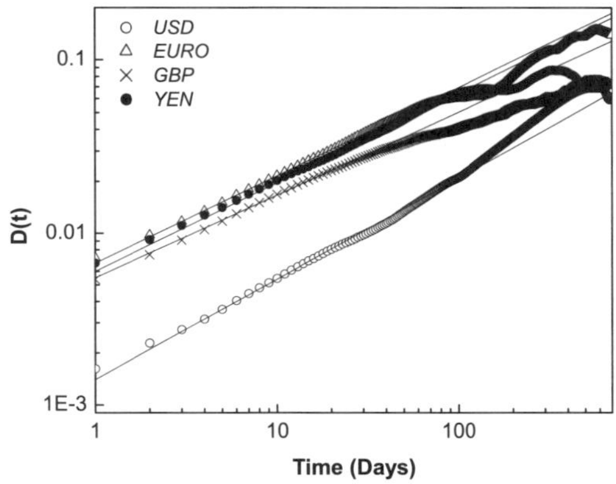

Fig. 2. SDA of the average daily exchange rate returns.

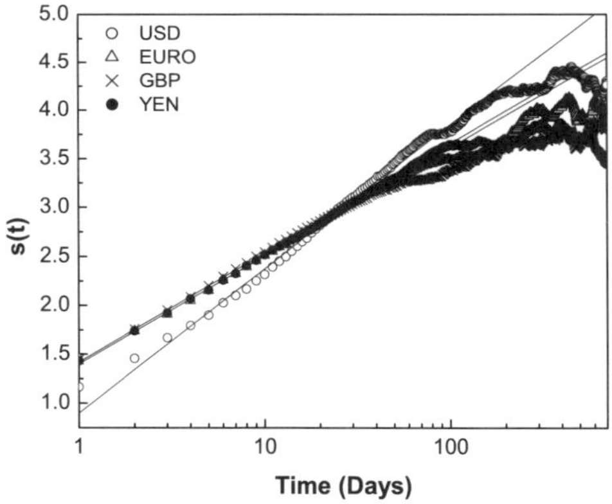

Fig. 3. DEA of the average daily exchange rate returns.

Table 1. Scaling exponents γ and δ obtained from SDA and DEA respectively for the average daily exchange rate returns of Indian Rupee versus the four foreign currencies.

Data	SDA exponent γ	DEA exponent δ
USD	0.59(0.02)	0.64(0.02)
EURO	0.50(0.02)	0.49(0.02)
GBP	0.48(0.02)	0.49(0.02)
YEN	0.51(0.02)	0.48(0.02)

exchange rate return data at random locations by adding noise of magnitude of the multiple of the standard deviation (std). We found that addition of noise of magnitude of five times of the std the scaling exponents did not change and the scaling behavior is retained by an addition of noise of magnitude of fifteen times of std. Which confirms the robustness of the scaling property of the INR-USD exchange rate return data. We have also analyzed the probability density distribution of the exchange rate returns. The distributions are fitted with Levy stable distribution, which is expressed in terms of its Fourier transform or characteristic function, $\varphi(q)$, where q is the Fourier transformed variable. The general form of the characteristic function of a Levy stable distribution is:

$$\ln(\varphi(q) = i\xi q - \eta|q|^{\alpha}\left[1 + i\beta\frac{q}{|q|}\tan\left(\frac{\pi}{2}\alpha\right)\right] \quad \text{for } [\alpha \neq 1] \tag{5}$$

where $\alpha \in$ (0,2] is an index of stability also called the tail index, $\beta \in$ [-1,1] is a skewness or asymmetry parameter, $\eta > 0$ is a scale parameter, and $\xi \in \Re$ is a location parameter which is also called mean. For Cauchy and Gaussian distribution, the values of α are equal to 1 and 2 respectively. A typical fit of the Levy stable distribution for INR-USD returns is shown in Fig. 4. The

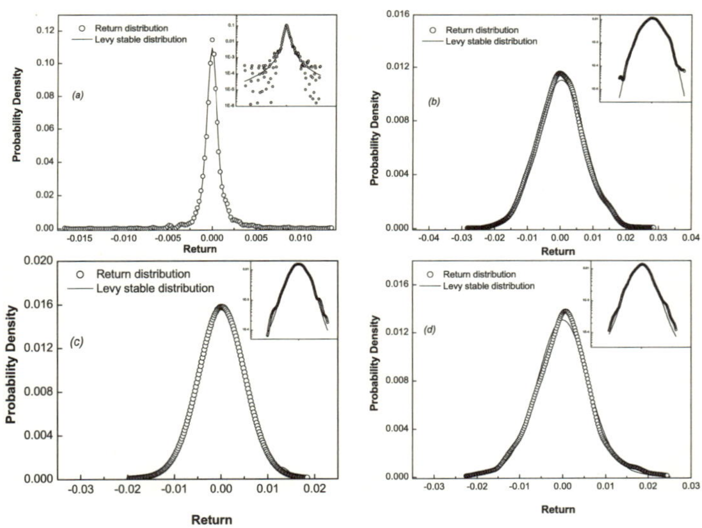

Fig. 4. Levy stable distribution fit of the (a) INR-USD (b) INR-EURO (c) INR-GBP and (d) INR-YEN average daily exchange rate return distributions. Insets in the figures show the plots in log-log scale (the return axis is shifted by 2 to show the negative tail).

parameters of the fitted Levy stable distribution for the average daily exchange rate returns of the four currencies are presented in Table 2.

Table 2. Parameters of the Levy stable distribution fit for the average daily exchange rate returns of the four currencies.

Data	α	β	η	ξ
USD	1.3307	0.1631	0.5376×10^{-3}	-0.4517×10^{-4}
EURO	1.9900	-0.9997	0.5117×10^{-2}	0.2436×10^{-3}
GBP	1.8860	-0.1005	0.3594×10^{-2}	0.1768×10^{-3}
YEN	1.8555	0.0713	0.4357×10^{-2}	0.2889×10^{-4}

From Table 2 it is seen that the value of α in case of INR-USD exchange rate is 1.3307 indicating the distribution is of Levy type but for the other cases values are close to the Gaussian limit 2. Which is also an indication of the randomness in those exchange rate returns.

3.2 Complexity

Recently Costa et al. [5] introduced a new method, Multiscale entropy (MSE) analysis for measuring the complexity of finite length time series. This method measures complexity taking into account the multiple time scales. This computational tool can be quite effectively used to quantify the complexity of a natural time series. The first multiple scale measurement of the complexity was proposed by Zhang [6]. Zhang's method was based on the Shannon entropy which requires a large number of almost noise free data. On the contrary, the MSE method uses Sample Entropy (SampEn) to quantify the regularity of finite length time series. SampEn is largely independent of the time series length when the total number of data points is larger than approximately 750 [7]. Thus MSE proved to be quite useful in analyzing the finite length time series over the Zhang's method. Recently MSE has been successfully applied to quantify the complexity of many Physiologic and Biological signals [5,8]. Here, we take the initiative to apply this novel method to quantify the complexity of a financial data.

The MSE method is based on the evaluation of SampEn on the multiple scales. The prescription of the MSE is: given a one-dimensional discrete time series, $\{x_1, ..., x_i, ..., x_N\}$, construct the consecutive coarse-grained time series, $\{y^{(\tau)}\}$, determined by the scale factor, τ, according to the equation:

$$y_j^{(\tau)} = 1/\tau \sum_{i=(j-1)\tau+1}^{j\tau} x_i \qquad (6)$$

where τ represents the scale factor and $1 \leq j \leq N/\tau$. The length of each coarse-grained time series is N/τ. For scale one, the coarse-grained time series is simply the original time series. Next we calculate the SampEn for each scale using the following method. Let $\{X_i\} = \{x_1, ..., x_i, ..., x_N\}$ be a time series of length N. $u_m(i) = \{x_i, x_{i+1}, ..., x_{i+m-1}\}$, $1 \leq i \leq N - m + 1$ be vectors of length m. Let $n_{im}(r)$ represent the number of vectors $u_m(j)$ within distance r of $u_m(i)$. $C_i^m(r) = n_{im}(r)/(N - m + 1)$ is the probability that any vector $u_m(j)$ is within r of $u_m(i)$. Approximate entropy (ApEn) introduced by Pincus [9] is defined as follows

$$\mathrm{ApEn} = \lim_{N \longrightarrow \infty} \varphi^m(r) - \varphi^{m+1}(r) \tag{7}$$

where

$$\varphi^m(r) = 1/(N - m + 1) \sum_{i=1}^{N-m+1} \ln C_i^m(r) \tag{8}$$

For finite length N the ApEn is estimated by the statistics

$$\mathrm{ApEn}(m, r, N) = \varphi^m(r) - \varphi^{m+1}(r) \tag{9}$$

Recently Richman and Moorman [7] have proposed a modification of the ApEn algorithm. They defined a new entropy, SampEn, which is given by

$$\mathrm{SampEn}(m, r, N) = - \ln \frac{U^{m+1}(r)}{U^m(r)} \tag{10}$$

In the case of SampEn the distance between two vectors is defined as the maximum absolute difference between their components and the vectors are not compared to themselves i.e. self-matches are ignored. For a given time series with N data points, only the first N-m vectors of length m, $u_m(i)$ are considered. This ensures the fact that, for $1 \leq i \leq N - m$, the vector $u_{m+1}(i)$ of length m+1 is also defined. Advantage of SampEn is that it is less dependent on time series length and is relatively consistent over broad range of possible r, m and N values. We have calculated SampEn for all the studied time series with the parameters m=2 and r= $0.15 \times$ SD (SD is the standard deviation of the original time series).
The result of the MSE analysis is shown in the Fig. 5. It can be seen that the the SampEn for the USD is smaller than the other three exchange rates at all the scales. This signifies the less complexity in the INR-USD exchange rate dynamics compared to the other three exchange rates.

The political development inside and outside a country affects its economy and the foreign exchange market. The cross currency volatility also influences a particular kind of exchange rate. The world economy experienced one of the worst shocks in the aftermath of September 11, 2001 events in the United States. Foreign exchange market in India also became volatile (shown in Fig.

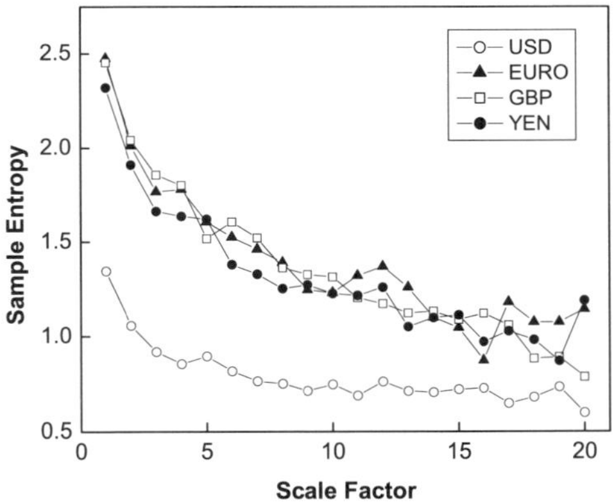

Fig. 5. MSE analysis of the daily average exchange rate returns

1a). Another large fluctuation in INR-USD exchange rate is observed around 31 March 2004. These fluctuations in the INR-USD exchange rate did not affect its robust scaling property. We argue this is due to the dissipation of the fluctuation in the vast economy of a country like India. The interacting elements provide a retarding path to the fluctuations in a financial market. As the number of interacting element increases the channel for the fluctuation dissipation gets broaden. USD is the most important foreign currency in the Indian economy. Hence, the number of interacting elements is more in the INR-USD exchange market. Possibly this is the reason behind the observed robustness of the scaling property in the INR-USD average daily exchange rate returns.On the other hand the complexity analysis revealed interesting result that the dynamics of the most important foreign exchnage market is the less complex. This result should be studied in detail.

4 Conclusion

The exchange rate management policy continues its focus on smoothing excessive volatility in the exchange rate with no fixed rate target, while allowing the underlying demand and supply conditions to determine the exchange rate movements over a period in an orderly way. Towards this end, the scaling analysis of the foreign exchange rate data is of prime importance. We have carried out extensive studies on the average daily exchange rate returns from Indian foreign exchange market. From the analyses we have found that the average daily exchange rate return of USD exhibits scaling and follows Levy Stable distribution. On the contrary, the average daily exchange rate returns of the other foreign currencies namely EURO, GBP and YEN do not follow scaling and they are found to obey Gaussian distribution.Moreover it is found that the complexity of the INR-USD foreign exchange market dynamics is less than the other three exchange rate.

References

1. Mantegna RN, Stanley HE (2000) An Introduction to Econophysics. Cambridge University Press, Cambridge
2. Reserve Bank of India, www.rbi.org.in
3. Scafetta N (2001) An Entropic Approach to the Analysis of Time Series. PhD Thesis, University of North Texas
4. Scafetta N, Hamilton P, Grigolini P (2001) Fractals 9:193
5. Costa M, Goldberger AL, Peng CK (2002) Phys. Rev. Lett. 89:068102
6. Zhang YC (1991) J. Phys. I (France) 1: 971
7. Richman JS, Moorman JR (2000) Am. J. Physiol. Heart Circ. Physiol. 278:H2039
8. Costa M, Goldberger AL, Peng CK (2005) Phys. Rev. E 71:021906
9. Pincus SM (1991) Proc. Natl. Acad. Sci. USA 88:2297

Noise Trading in an Emerging Market: Evidence and Analysis

Debasis Bagchi

The Institute of Cost and Works Accountants of India, 12. Sudder Street, Kolkata 700 016, India
dbagchi2002@yahoo.co.in

Summary. We document influence of nascent stock market noises on the stock price movement in an emerging market by examining behavior of three noise indicators given by spread of high and low quotes of the stock, spread of opening and previous closing prices and turnover. The results indicate that investors indeed use these noises but to a limited extent to predict future prices. We also find that there is no significant difference of behavior of noise traders during different economic conditions of the market, viz., falling and rising market conditions.

1 Introduction

The movements of stock prices have attracted fascination of the investors and academics alike. While investors are overtly interested to make profits, academics are concerned with how prices are determined and whether future prices can be predicted. Hitherto the most developed theory on financial market and understanding movements of stock prices is based on market efficiency or Efficient Market Hypothesis (EMH). An efficient market is defined as a market where large numbers of rational profit-maximisers are actively competing to predict future market prices of individual securities and where information is freely available to all participants. In such an ensued competition actual prices of the individual securities reflect the impact of information on past and present events and also market expectation on future events. An important implication of existence of such a market is that the future market price of the security will be dependant on future information, which by nature is unpredictable and, therefore, the future market price will also be unpredictable. Initially during 1970s evidences are overwhelmingly in favor of random nature of the stock prices, upholding the hypothesis. It was subsequently challenged in 1980s on the basis of evidences that pointed out that apart from information that are reflected in the security prices, there are several other variables that can be relied upon to predict the future prices of the securities. During 1990s researchers looked beyond the scope of existing theory towards finding an alternate market efficiency mechanism.

2 Motivation

Evidences against EMH are varied and widespread. However, investigation of the behavior of short-term investors is not sought to be evaluated per se. Lo and McKinley (1999) investigated short-run stock prices and found non-zero serial correlation in the process, with the existence of a large number successive moves in the same direction. However, this study did not investigate how the investors use nascent market generated information or nascent noise for formulating a mechanism for future price discovery. The researchers have also documented several indicators that are found to have predictive ability, but attempts are hardly made to investigate behavior of investors when nascent market information is released to them. In addition the studies are mainly restricted to developed market and research relating to emerging market is rare. Since emerging market is viewed as a less efficient market, an examination of the behavior of noise traders in such market is important. It is expected that being less efficient, the future price determination in such a market would be driven by a considerable extent on noise trading rather than on fundamental information.

A study on the short-term behavior of the investors as to how they tend to discover the profit opportunity on the basis of nascent market generated information is important at this stage. It is likely that the investors would formulate their decision rules based on some indicators, built up on these nascent information. Since they are not related to the fundamental performance of the firms, they are called noises. This study looks into the aspect of identifying such indicators. We accordingly seek to investigate : (1) whether the investors use market generated nascent information to formulate their investment decision and (2) how the investors use the same information during different economic condition of the stock market, that is, during falling market condition and during rising market condition. An investigation of the behavior of the investors during these two economically opposite period is likely to reveal nature of investors behavior on nascent market information.

Globally, one of the important emerging markets is the Indian Stock Market, which has the distinction of witnessing falling and rising market conditions in a short period of time. Such market trends provide us an opportunity to study the behavior of noise traders during two economically opposite conditions. Accordingly, this research focuses on evaluating the behavior of noise traders in Indian Stock Market.

3 Methodology

A multiple regression model is used to evaluate the effect of market generated nascent information on the price of individual security. The indicators are built up on the market information relating to the security. The business newspapers usually publish stock market data relating to the prices of individual security such as, previous closing price, opening, high, low and closing price of the security. The investors are likely to be influenced by intra-day volatility as given by the high and low prices, since the difference between high and low prices would be construed as an opportunity to make a short-term trading profit. The proxy for overnight volatility is given by the difference between previous closing price and opening price. The investors in order to assess the transmitted volatility of the security are expected to use these information in their effort to build up their own model for price discovery.

The direction of price movement is likely to be used to formulate their own trading strategy. A high positive rise may induce the investors to invest while a high negative slide will, in turn, induce the investors to withdraw. The proxy of the overall volatility is given by the days turnover of the stock. A high turnover is an indication of investors choice. Since, we want to understand influence of nascent noises on the stock price movement, we use the price of the security as dependant variable while the indicators are used as independent variables in the regression model. The absolute values of the stock prices are converted to their log values. In addition two indicators, viz., spread of high and low and the turnover are converted to their logarithmic values. Because of possibility of negative quantity the spread of opening and previous closing price is not converted to its log. Since the investors would use these noises to predict future prices of the stocks, on a short-term time horizon, these indicators are calculated on the basis of previous days prices, i.e., the most nascent information.

The following multiple regression model is used in our study:

$$Y = \beta_0 + \beta_1 x_{1i} + \beta_2 x_{2i} + \beta_3 x_{3i} + \epsilon \quad (1)$$

Subject to usual assumption like, $E(\epsilon) = 0$ and ϵ re normally distributed and there are no linear dependencies in the explanatory variables, etc., and where:

Y = log of stock price

x_{1i} = log (Spread of high minus low quotes of previous day), (HL)

x_{2i} = Spread of opening minus previous closing price of previous day, (OP)

x_{3i} = log (Turnover of previous day) (T)

β_0, β_1, β_2, β_3 are regression coefficients and ϵ is the error term.

We test the model for a possible existence of unit root. The Augmented Dickey Fuller test rejects the hypothesis for existence of unit root. We also make the results White heteroskedasticity-consistent for standard errors and covariance.

4 Data

The firms which are fifty in number that are used to construct the stock index of National Stock Exchange of India, known as Nifty, are selected for the study. Since the study is also to investigate the nature of behavior during the distressed as well as flourishing market conditions, two relevant time periods are accordingly selected which reflect such conditions. First, the period between November 1, 1999 and October 31, 2000 is selected for distressed market condition. During this period, the market behaved in a distressed manner, with around 30% drop in market capitalization as well as the index points as the stock prices of the Information Technology companies in India crashed. Second, we also choose the period between November 1, 2004 and October 31, 2005, to reflect flourishing market condition. During the above period the stock market in India has exhibited an unprecedented surge in the share prices leading to phenomenal rise of more than 50% of index point. All the observations are on daily basis. Identical calendar period is selected to reduce the bias that may enter in our investigation due to various observed effects, like January effect, volatility effect, etc. For the purpose of making comparative study, we have to reduce the sample firms to 37, since the remaining 13 firms are rejected due to the fact these firms do not constitute the Nifty index during both the periods. Suitable adjustments are made to accommodate bonus shares, new share issues, stock splits etc. The observations are recorded on daily basis.

5 Results

The results of the regression analysis are given in the Table 5 for the period 1999-2000 (distressed period) and in the Table 2 for the period 2004- 05 (flourishing period). The adjusted R^2 is found to be modest to low which underlies that the cross-sectional variations of price fluctuations are explained low to moderate extent by the variables.

Our basic aim is to find out whether the noises, as given by the variables selected for our study are, in fact, used by the investors to formulate their short term investment decisions. The study also aims to provide an explanation of the investors behavior in two opposite economic conditions, viz., falling market condition and rising market condition. It is observed from the regression analysis that β_0 has a large statistically significant value with respect to those of three noise variables. Since, β_0 signifies factors other than the explanatory variables of the regression equation, it follows that investors to a large extent depend on other information rather than on the noise. Nevertheless, it is observed that for a large number of sample firms noise plays an important limited role for price discovery. Since, we experiment with previous days price variations, it shows the speed through which noise is absorbed in the market and the accompanying effect that it brings in to the capital market. Our primary aim is to find out how the noise, mistakenly regarded as information by the investors, influences the stock price behavior. Individually, the variables show different characteristics for each firm. Only in six cases, the impact of these variables is found to be nearly identical on the firms over the two time periods. On the other hand, each individual variable exhibits

Table 1. Regression Results – Period: November 1, 1999 to October 31, 2000 (Distressed Period)

Firms' Name	β_0	t	β_1	t	β_2	t	β_3	t	R^2	Adj R^2
ABB	2.134	98.646*	.09141	3.824*	-.000187	-.232	.085	5.224*	.343	.335
ACC	1.297	16.393*	.144	4.129*	.004701	2.095	.213	7.253*	.552	.546
B. Auto	2.486	177.52*	-.0501	-4.76*	-.000882	-1.004	.049	6.848*	.162	.152
BHEL	2.253	32.17*	.334	9.73*	.004582	1.979	-.134	-4.75*	.287	.278
BPCL	2.242	55.15*	.274	10.91*	-.001084	1.010	-.0814	-4.31*	.333	.325
Britannia	2.788	8.60*	-.00356	-.161	-.000454	1.248	.0211	1.321	.016	.004
Cipla	2.646	42.80*	.229	8.97*	.0085	3.861*	-.0186	-.799	.306	.298
Colgate	2.000	92.89*	-.0551	-2.79*	.00051	.352	.132	10.88*	377	.369
Dabur	2.434	60.81*	.230	10.15*	-.00021	-.796	.057	2.71*	.437	.430
Dr Rddy	3.054	70.81*	-.05482	2.928*	.00021	1.640	-.012	-.729	.046	.035
GAIL	1.847	75.25*	-.000532	0.022	.0059	2.087	-.009	-.547	.047	.015
Grasim	2.098	42.38*	.136	4.31*	.00166	1.956	.086	3.404*	.271	.262
Guj Amb	2.113	23.46*	.355	9.91*	-.00041	-.919	-.042	-1.043	.329	.321
ITC	2.562	51.78*	-.0229	-1.164	.000966	3.282*	.09	5.993*	.169	.159
HCL Tec	2.676	31.91*	.101	3.49*	.00061	4.875*	.0807	2.601*	.252	.240
HDFC	2.018	45.72*	.151	4.85*	.0028	1.994	.068	2.847*	.276	.267
Hero	2.900	128.53*	.02474	1.702	.00025	1.874	.027	2.828*	.092	.080
HLL	3.101	24.92*	.803	38.68*	-.00009	-1.249	-.405	-9.04*	.861	.860
HPCL	1.961	46.71*	.103	4.554*	.00409	2.276	.037	1.962	.188	.178
Ind Hotl	2.221	110.97*	.119	6.071*	-.00216	-2.747*	.039	2.968*	.284	.275
IPCL	1.299	29.20*	-.001869	.053	.007	-1.66	.223	10.02*	.440	.433
Infosys	4.343	67.60*	-.07198	5.179*	.00002	2.433	-.138	-9.97*	.336	.328
MTNL	2.404	34.57*	.411	16.30*	.0042	3.99*	-.151	-6.31*	.577	.572
M & M	2.202	31.08*	.289	11.26*	-.0044	-4.7*	-.038	-1.309	.400	.393
ONGC	1.948	78.59*	.112	3.652*	-.00097	-.507	.097	5.154*	.252	.242
Ort Bnk	1.422	137.47*	.004996	.321	-.00055	-.149	.106	12.69*	.592	.587
Ranbaxy	2.242	84.62*	-.101	-5.35*	-.00003	-.128	.209	22.15*	.733	.730
Reliance	2.204	31.97*	-.0122	-.513	-.00062	-.527	.0069	3.56*	.069	.058
Satyam	2.863	16.34*	.744	29.79*	-.00022	3.05*	-.241	-6.52*	.793	.791
SAIL	.633	15.01*	.114	4.466*	.015	.896	.160	8.885*	.677	.673
SUN	1.930	25.17*	.524	14.90*	.00028	3.267*	.036	1.009	.583	.578
T. Chem	1.360	82.28*	-.0574	-3.34*	.0052	1.946	.182	17.06*	.608	.604
TISCO	1.648	40.14*	-.0474	-2.024	.0016	.729	.145	9.059*	.357	.349
Tata Tea	2.088	38.54*	.309	10.15*	.0012	1.521	.038	1.376	.494	.488
VSNL	2.484	38.69*	.414	14.36*	.00025	1.521	-.065	-2.94*	.475	.469
WIPRO	2.878	36.28*	.149	5.233*	.00025	3.374*	.0645	2.814*	.235	.225
Zee	2.427	10.03*	.693	20.98*	.00005	1.498	-.160	-3.17*	.652	.648

* significant at 1% level

Table 2. Regression Results Period: November 1, 2004 to October 31, 2005 (Rising Period)

Firms' Name	β_0	t	β_1	t	β_2	t	β_3	t	R^2	Adj R^2
ABB	2.718	62.453*	.03331	1.274	-.00054	-1.65	.133	7.772*	.271	.262
ACC	2.630	65.012*	.165	9.301*	.00075	1.17	-.0593	-4.65*	.261	.252
B. Auto	2.642	62.455*	.120	5.727*	-.00041	1.519	.0826	5.796*	.329	.321
BHEL	2.969	55.88*	.226	10.99*	.00101	2.823*	-.107	-5.59*	.329	.321
BPCL	2.443	126.70*	.00025	.022	.00033	.731	.0464	6.744*	.199	.189
Britannia	2.878	151.53*	.07454	5.964*	.0038	.163	-.0043	-.487	.142	.131
Cipla	2.467	66.694*	.08216	5.435*	-.00068	.978	-.0204	-1.571	.114	.103
Colgate	2.062	89.031*	.03697	1.781	-.00087	-.650	.0938	8.564*	.338	.330
Dabur	1.831	41.647*	.104	3.807*	-.0031	-1.36	.0698	3.541*	.210	.201
Dr Rddy	2.819	130.79*	.0033	2.961*	-.000026	-.108	.00735	.884	.066	.055
GAIL	2.692	16.89*	.878	26.32*	.02473	4.955*	-.312	-6.30*	.141	.741
Grasim	2.276	105.38*	.0373	3.764*	-.000073	-.084	.00157	2.091	.141	.130
Guj Amb	3.622	25.236*	.766	24.84*	.0004	2.010	-.451	-5.66*	.775	.773
ITC	3.003	121.21*	.01675	1.408	-.00022	-1.337	.0188	2.199	.051	.040
HCL Tec	2.509	77.693*	.0809	5.486*	-.00052	-1.301	-.0065	-.589	.163	.153
HDFC	2.578	75.342*	.0597	4.386*	-.00017	-.702	.0335	2.934*	.153	.143
Hero	2.694	72.731*	.113	6.962*	.00086	2.095	-.024	-1.809	.175	.165
HLL	1.928	53.818*	.0458	3.210*	.0014	3.210*	.06356	5.465*	.280	.271
HPCL	2.310	114.44*	.003455	.324	-.00018	-367	.0614	8.633*	.307	.298
Ind Hotl	2.594	110.21*	.04463	2.505	.0011	3.206*	.0596	6.872*	.289	.280
IPCL	2.007	92.402*	-.00498	-.416	-.0004	-.256	.0743	9.677*	.407	.400
Infosys	3.365	65.760*	.02896	2.270	.000005	.048	-.0178	-1.359	.021	.009
MTNL	1.872	81.502*	-.041	-3.29*	.0112	.990	.0859	10.09*	.330	.322
M & M	2.479	37.634*	.0929	3.899*	.000425	2.133	.0327	1.595	.104	.093
ONGC	2.903	75.039*	.0715	6.992*	-.00048	-2.60*	-.0113	-1.085	.178	.168
Ort Bnk	2.250	94.073*	.0109	.765	.0026	3.536*	.0583	6.408*	.277	.269
Ranbaxy	3.346	33.192*	.429	9.338*	.00075	3.065*	-.274	-8.39*	.302	.294
Reliance	2.284	28.5*	.0333	1.485	.00055	.914	.100	4.874*	.188	.178
Satyam	2.422	30.89*	.0839	4.90*	.00043	.464	.0336	1.600	.131	.120
SAIL	1.594	39.49*	.0436	3.098*	-.0049	-1.442	.0403	3.699*	.172	.161
SUN	2.635	89.977*	.0821	5.741*	.00085	2.558*	-.00167	-.135	.154	.144
T. Chem	2.021	93.976*	-.0003	-.020	.00154	1.392	.0698	7.815*	.299	.290
TISCO	2.263	34.567*	.0242	1.687	-.00025	-.677	.0673	3.966*	.124	.113
Tata Tea	2.477	49.799*	.197	7.924*	-.00166	-1.761	.0204	1.078	.265	.256
VSNL	1.947	140.60*	-.083	-4.84*	.00101	1.164	.171	26.28*	.832	.830
WIPRO	2.812	25.859*	.197	5.968*	.00074	2.766*	-.0725	-2.104	.139	.128
Zee	1.930	81.901*	-2.23	-1.574	.0001	.104	.0851	9.449*	.366	.358

* significant at 1% level

approximately an alike behavior for a cluster of firms. The number of firms forming this cluster is found to be highest for the HL variable, while it is least for the variable OP. The table given below exhibits some of the characteristics of the three variables selected variables or noises.

	β_1		β_2		β_3	
	Falling	Rising	Falling	Rising	Falling	Rising
Minimum	(-)0.101	(-)0.083	(-)0.0044	(-)0.00048	(-)0.405	(-)0.451
Maximum	0.803	0.878	0.0085	0.02473	0.223	0.171
Mean	0.222	0.1595	0.00089	0.00363	0.0266	0.0084
Std. Dev.	0.242	0.2273	0.0037	0.00795	0.1492	0.1531
Significant Cases (No.)	29	24	9	9	28	23
(-) Coefficient (No.)	6	2	3	1	8	5

It is observed that none of the variables shows any widely different behavior between falling and rising market conditions. Contrary to the general argument that during distressed market condition, the market is more efficient and investors would act on the basis of fundamental information predominantly than on noise, it is found that investors use noise in a differentially higher order in the falling market conditions than in the rising market. A plausible explanation could be that, in a falling market the investors are likely to be panicky which leads them to use whatever information they have, while during rising market, there has been little panic and that leads to a sublime condition of using more of fundamental information by the investors. In addition, unlike in rising market where investors depend more on long-term investment decision, during falling market, majority of the investors are induced to take short-term investment decision till the equilibrium is reached.

The variable HL shows marginally different characteristics during falling and rising market conditions. As expected in majority of cases the coefficient is found to be positive. It strengthens our argument that nascent noise in fact contributes to the process of future price discovery.

The overnight volatility is found to be least useful for the noise traders, since in only around 25% of the sample cases the variable OP is found to be statistically significant. In addition the impact of the variable on price discovery process is very low. However, the impact is found to be higher in rising market condition compared to falling market condition a fact that is well supported by other research findings. The turnover variable (T) shows a similar behavior with respect to other two variables. The impact is differentially more in falling market than in rising market. The observed behavior is partially explained by our earlier argument on falling market inducement for taking short-term investment decisions and the results suggest that in the process the efforts of the investors for discovery of future price are supplemented. In sum, we document that in spite of an unknown factor playing a dominant role, the nascent noises are, in fact, to limited extent used by the investors to predict future prices and to formulate their decisions rules. The results show that intra-day volatility noise as given by the variable HL is most used, while the investors least use overnight volatility measure OP. The investigation of the behavior of the noise traders during falling and rising phases of the market does not reveal any remarkable difference of behavior. Such a phenomenon establishes that in an emerging market during economically opposite phases the efficiency of the market remains the same.

References

1. Black F (1986) Noise. Journal of Finance 41:529–544
2. Fama E (1970) Efficient Capital Markets:A review of Theory and Empirical Work, Journal of Finance 25:383–417
3. Lo A, Mackinlay C (1999) A Non-Random Walk Down Wall Street. Princeton, NJ, Princeton University Press.

How Random is the Walk: Efficiency of Indian Stock and Futures Markets

Udayan Kumar Basu

Future Business School, Kolkata, India
udayan_basu@yahoo.co.in

1 Prelude

Time series of prices of stock and its rates of return has been one of the major areas of study in Econophysics. The price of a stock depends on a number of factors as well as information related thereto, and how quickly and effectively the price of a stock assimilates all such information decides the efficiency of the stock market. Instead of individual stocks, people often study the behaviour of stock indices to get a feel of the market as a whole, and the outcomes of such studies for the Dow Jones Industrial Average (DJIA), the Nasdaq Index and the S & P 500 Index have been listed in a number of articles. In this context, it has also been argued that for a market to be considered sufficiently liquid, correlation between successive price movements and rates of return should be insignificant, because any significant correlation would lead to an arbitrage opportunity that is expected to be rapidly exploited and thus washed out. The residual correlations are those little enough not to be profitable for strategies due to imperfect market conditions. Unless transaction costs or slippages or any other impediment exists, leading to some transactional inefficiency, arbitrages would take place to bring back the markets to a stage of insignificant correlations [1, 2].

With substantial foreign inflows into the Indian stock markets during the last few years, the volumes and market capitalisations at the bourses have literally soared, with the stock indices scaling all time highs. Of late the Japanese fund managers have also flocked to India and the inflow of funds from Foreign Institutional Investors (FIIs) has exceeded USD 10 Billion in 2005. The markets witnessed considerable buoyancy during the period under review. Although the Mumbai Stock Exchange (BSE) is the oldest exchange of India having a large number of listed companies, the National Stock Exchange (NSE) has stolen a march over it in so far as the total volume of transactions is concerned. Besides, the futures market has already overtaken the cash market in the overall volume of transactions. In 2004, the volume in the futures market was two and a half times that in the cash market and in 2005 the former has reportedly exceeded three times the latter. As for the futures transactions also, the NSE once again hogs the limelight.

In order to exclude the possible influence of environmental factors on any particular company or industry during a particular period of time, it may be better to study the market index and the index futures, such as the Nifty and the Nifty Futures, instead of individual stocks and stock futures. In other words, the stock index and the stock index futures are better candidates to provide unbiased estimates for the market movements. Over a period of eighteen months between April 2004 and September 2005, the behaviour of the Nifty and Nifty futures has been gone into. The prices and the rates of return of the Nifty over the above eighteen months have been examined and the serial correlations have been worked out. The serial correlations for Nifty Futures and also for the basis (difference between Stock Index Price and the Stock Index Futures Price) have been analysed to understand the characteristics of the Stock and the Futures markets and the possible inter market correlations.

2 Efficient market hypothesis and serial correlation; market beta

Eugene F. Fama of the University of Chicago was the one to propose the Efficient Market Hypothesis. This suggests that in an ideal market all relevant information is already priced into any security today. In other words, yesterdays changes do not influence todays changes and nor do todays changes influence tomorrows. This hypothesis of information efficiency of stock markets would imply that each price change is independent of the last one. If the price or the rate of return of a security has increased during a certain interval of time, there is no way to predict whether in the next interval the same parameter will increase further or decrease. This aspect was explored even earlier by Bachelier, who did not stop at the theory but also tested his equations against real prices of Options and Futures contracts. The theory seemed to work [3]. As for the various kinds of correlations, the strongest ones are most often the short term ones between periods close together and the weakest are those between periods far apart. As a matter of fact, if all the correlations, from short term to long term, are plotted one normally ends up with a rapidly falling curve. So, while looking for correlations in the markets, one would normally concentrate on the short term ones. Investors in the Indian markets are primarily concerned about whether an increase in stock price or rate of return during a certain day would lead to a further increase or a decline during the next day. In other words, the one-day correlation among daily closing or settlement prices appears to be the most relevant yardstick for analysis.

A perfectly efficient market would require all correlation functions for non-zero time lags to be zero. However, such a market is an impossibility in practice, for trading in a perfectly efficient market would be hardly of any use.

Bachelier & Samuelson argued that price variations in time are not independent of the actions of the traders, on the contrary, it results from them. According to them, any advantageous information that may lead to a profit opportunity is quickly eliminated by the feedback that their action has on the prices. If such feedback action occurs instantaneously, as in an idealized world of idealized frictionless markets and costless trading, then prices must always fully reflect all available information and no profit can be garnered from information based trading since such profits have already been captured.

If all informed traders agree at a certain point of time that buying is the best policy, in the absence of noise traders, they would not find any counterparty and there would be no trade. Thus, the stock market requires the existence of some noise, howsoever small, which provides liquidity to the market.

Even the weak form of the Efficient Market Hypothesis, popularly known as the Random Walk Theory, directly repudiates Technical Analysis, and a series of historical stock prices or rates of return cannot be of any assistance in predicting stock prices or rate of return in future. The semi-strong form militates against the spirit of Fundamental Analysis, based on analysis of published data on company, group, industry and economy.

While the technical analysts analyse various charts and look for patterns like Head and Shoulder etc., fundamental analysts study EPS, outlooks for the economy, industry, group and company in order to get a feel of the future prospects for a corporate and its common stock. If the markets are truly efficient, both the technical and fundamental analyses will fail to give any indication as to the possible future movement of the price of any stock. Even insider information may not lead to extraordinary returns on a long term basis. In other words, no activity would, on the average, lead to a long term gain exceeding the average long term return of the stock market. It is cheaper and safer to ride with the market.

So, any success with Fundamental and Technical Analyses would imply that the market may be efficient, but still requires some time before the impact of all data and inputs is felt completely on security prices. As the market becomes more and more efficient, the time available for market players to act will decrease and they have to become smarter in order to reap the benefit of their analysis. Since the random walk theory is interested in testing the independence between successive price changes, correlation tests are particularly appropriate. In particular, one would be interested to examine whether price changes during a period $t+1$ are correlated to price changes in the preceding period t. If on an average, a price rise (fall) during period t is to be followed by a price rise (fall) in period $t+1$, then correlation coefficient is large and positive (close to $+1$). On the other hand, if the price declines (increases) on the average in period $t+1$ after a rise (fall) in period t, the correlation coefficient is large and negative (close to 1). For an efficient market, the increments have random signs and have amplitudes distributed according to the so called

Gaussian distribution. After all, information and efficiency seem to go along with randomness.

Let us suppose that we have an infinite link of identical beads in a uniform chain. As we try to colour the beads differently, each colour standing for some codified information, the uniformity or the repetitiveness of the beads will disappear. If we wish to work out the shortest algorithm to contain the entire information, all the useful beads are likely to have different colours and the correlation among them will be completely lost. The un-coded beads will however, still display a long term order. In other words, greater information content is likely to be accompanied with greater randomness in the structure.

Liquidity is about the ability to transact at a price, which is very close to the current market price. For instance, a stock is considered liquid if one can buy some shares at around Rs.320.05 and sell at around Rs. 319.95, when the market price is ruling at Rs. 320. A liquid stock has very tight bid-ask spread. Two measures are normally used to measure liquidity. These are turnover (total value traded) in the stock market as a ratio of (i) Stock Market Capitalisation & (ii) GDP. The first measure is also called the Turnover ratio. A high turnover ratio is normally associated with a low transaction cost. Liquidity is a pre-requisite for investors to carry on transactions and earn profits. An absolutely efficient market without noise traders may lead to complete drying up of the market and total lack of liquidity. Perfect liquidity, on the other hand, may blunt the market efficiency. Presence of some noise traders is thus a sine-qua-non for ensuring a degree of liquidity for the markets.

Markets can exhibit dependence without correlation. For this, we have to distinguish between the magnitude and the direction of price changes. Suppose that the direction is uncorrelated with the past. In other words, the fact that prices fell yesterday does not make them more likely to fall or rise today. Still, it is possible for the absolute changes to be dependent. A 10% fall yesterday may well increase the odds of another 10 percent move today although it tells nothing about whether the price will move up or down. In such a situation, there is a strong dependence but no correlation. Large price fluctuations would tend to be followed by more of large fluctuations, positive or negative and small changes tend to be followed by more of small changes. This leads to clustering of volatility.

Runs tests ignore the absolute values of the number in the series and observe only their signs. In such a study, the researchers merely count the number of runs- i.e. consecutive sequence of signs in the same direction. Random Walk Theory would suggest that previous changes of prices or rate of return are useless in predicting changes in future prices or rates of return. In other words, successive price changes are independent. Given a series of stock price changes, each price change is designated by a plus (+) if it represents an increase or a minus (−) if it represents a decrease and a run is said to occur when there is no difference between the signs of two successive changes. When the sign of change differs from the previous one, the run ends and a new run is said to begin.

One way to test the randomness of stock prices, the rates of return on stocks is to focus on their serial or autocorrelations. Such changes are serially independent if the autocorrelations are negligible. Numerous studies, employing different stocks, different time lags and different time periods have been conducted to detect serial correlations. These studies are concerned with short term (daily, weekly etc.) and not long term correlations, because in the long term stock prices tend to move upward. Besides, there may be the so-called long range correlations leading to predictability which explains the phenomenon of market crash. The hypothesis of Didier Sornette is that stock market crashes are caused by the slow build-up of long-range correlations leading to a global cooperative behaviour of the market and eventually leading to a collapse in a short critical time interval (local self-reinforcing imitation among traders).

For computing the rate of return on stock or a stock index, we may write down the price at time $t+1$ as

$$S(t+1) = u(t)s(t) \quad \text{for} \quad t = 0, 1, \ldots, N-1.$$

Here the quantities $u(t)$ for $t = 0, 1, 2, \ldots, N-1$ are mutually independent random variables. The variable u(t) defines the relative changes in prices between time t and $t+1$. The relative change is $S(t+1)/S(t)$, which is independent of the overall magnitude of $S(t)$ and also the unit of price. Then, $\ln S(t+1) = \ln S(t) + \ln u(t)$. Let $w(t) = \ln u(t)$; or, $u(t) = \exp[w(t)]$. Each of the variables $u(t)$ is said to be a lognormal random variable since its logarithm is an actual normal variable. The price distribution of stocks and stock indices are actually quite close to lognormal.

The rates of return on stocks are not symmetric because stock prices may have unlimited upward movements while the downward movements are restricted due to the fact that prices of stocks can never be negative. But, if we define the rate of return as the logarithm of the ratio of successive stock prices, this is distributed symmetrically around zero.

If the stock prices for N successive time intervals are noticed, the price movements or rates of return on stocks for $(N-1)$ successive time intervals can be worked out. Considering only the signs of such price movements or rates of return, we can record the sequence of pluses and minuses. Based on these, we can work out the Kendall Rank Correlation connecting successive price movements or rates of return. In that case, the study will reveal the correlation and not the dependence among successive stock prices and rates of return.

Beta for a security or a portfolio of securities is defined as its covariance with the market portfolio (market index) divided by the variance of the market portfolio. In other words, $\beta_i = \frac{\sigma_{i,M}}{\sigma_M^2}$ where M stands for the market portfolio. It is a measure of the systematic risk associated with the security or portfolio of securities. It also specifies the percentage change in the value of the security/portfolio for a one percent change in the market portfolio.

In other words, change in the price of the security/portfolio consequent upon a change in the market is equal to its beta times the change in the market portfolio. Or, the hedge ratio for a futures contract is inverse of its beta.

3 National stock exchange (NSE); Nifty and Nifty Futures

Among the stock exchanges in India the NSE is marked with the largest volume of transactions and is considered the most liquid. The S & P CNX Nifty is an index based on 50 stocks, which are chosen based on their volumes of transaction, market capitalization and impact cost. This index, which was introduced in 1995 (closing price as on 03/11/95=1000) is currently close to 3000. The market impact cost is a measure of the liquidity of the market. It reflects the costs faced when actually trading an index. The market impact cost on a trade of 3 million rupees of the full Nifty works out to about 0.2%. This means that if Nifty is at 1000, a buy order goes through at 1002 i.e. $1000+0.002 * 1000 = 1002$ and a sell order gets 998 only. The Nifty has registered substantial movements during the eighteen months under consideration.

Nifty Futures is one of the best derivative securities introduced in the Indian securities market. It was introduced in 2000 and trades on the Futures & Options segment of NSE. The contract multiplier is currently 100 although it was 200 to start with. This means the value of one Futures contract is rupees one hundred times the futures quotation. The current price of this index futures is close to Rs. 3 lacs. The fair value of any futures (F) is normally linked to the corresponding spot price (S) by the equation:

$$F = S(1 + r - q)T,$$

where r = cost of financing,
q = expected dividend yield,
T = holding period.
In case of continuous compounding, we can write $F = S\exp[(r - q)T]$

The basis is defined as the difference between the spot price and the futures price. Normally, for a positive cost of carry, basis is negative and equal to the cost of carry. For a negative cost of carry, basis is positive and equal to the cost of carry. Cost of carry = cost of holding a stock less impact of any dividend yield. Any departure from this scenario will be adjusted through arbitrage between the Cash Market and the Futures Market. If futures has a value higher than its fair value (F above), one would buy in the cash market and sell in the futures market to reap the arbitrage opportunity. On the other hand, if futures has a value lower than its fair value, reverse arbitrage would be resorted to by the market players by selling in the cash market and buying in the futures market. At the time of maturity of the futures contracts, its value equals the spot price in order that no instant, risk-free profit may result from any arbitrage operation. In other words, at the time of maturity, the futures price converges with the spot price and the basis consequently becomes zero.

At any point of time, there are three Nifty Futures Contracts in existence. The first one, which matures in the same month, is called the near month contract. The other two are called the next month and the far month contracts. These contracts mature on the last Thursdays of the months to which they relate. On the next working day, another new contract is floated which then becomes the new far month contract. The erstwhile far month and mid month contracts now become the new mid month and near month contracts respectively. Over the eighteen month period under study, therefore, we come across 20 futures contracts. However, sixteen of them only survive for 3 months and we restrict our focus to these sixteen contracts only (Number 3,4,...,18) [4].

4 Review of earlier work

Sharma and Kennedy (1977) used run tests to study the behaviour of BSE stocks and concluded that they obey a random walk.

Kulkarni (1978) investigated the weekly RBI Stock price indices for Bombay, Calcutta, Delhi and Madras stock exchanges and rejected the random walk hypothesis.

Yalawar (1988) studied the month-end closing prices of 122 stocks listed on BSE during 1963 82. He used runs tests & Spearmans rank correlation. He found Lag 1 correlations were significant for a number of stocks (21 out of 122).

Anand Pandey of IGIDR (Indira Gandhi Institute of Development Research), Mumbai computed autocorrelations of stock prices (CNX Nifty, CNX Defty, CNX Nifty Junior) for the period January 1996 to June 2002 on the basis of daily closing prices. He concluded that market is not very efficient and it is possible to pick up undervalued stocks to reap benefits. A stock is said to be undervalued when the market price is less than its intrinsic value. The intrinsic value is the sum of the discounted values of all future inflows (perpetual) from the stock and is supposed to reflect fundamental value.

R. N. Agarwal of Delhi University also studied the efficiency of stock markets. He argued that:

(a) Following liberalizations of the financial markets, good amount of inflow from FII (foreign institutional investors) takes place FIIs come to India in order to maximize their rate of return, at the same time diversifying their portfolio further (International Diversification)
(b) CAPM presumes market efficiency. In an efficient market the expected real return on stock should equal the real interest on risk-less assets and the premium on risk taking. A market is allocationally efficient when prices adjust so that the risk-adjusted rate of return is equal for all investors.

(c) If Capital Markets are efficient and the rate of return on any asset is a fair game, then, on the average, across a large number of samples, the expected return on an asset equals its actual return.

Results for the period 1988-91 suggest that pricing in Indian Stock Markets is not consistent with the risk-return parity postulated by CAPM.

Arusha Cooray of the University of Tasmania (School of Economics) investigated in his paper whether the stock markets of South Asia (India, Pakistan, Sri Lanka and Bangladesh) followed a random walk. Employing, inter alia, autocorrelation tests and spectral analysis and using monthly data over the period from 1996.1 to 2003.10, he concluded that all these markets exhibit a random walk indicating weak form efficiency in the stock markets of South Asia. He also showed that stock price shocks in India have greater effects on Pakistan and Sri Lanka than on Bangladesh [5–7].

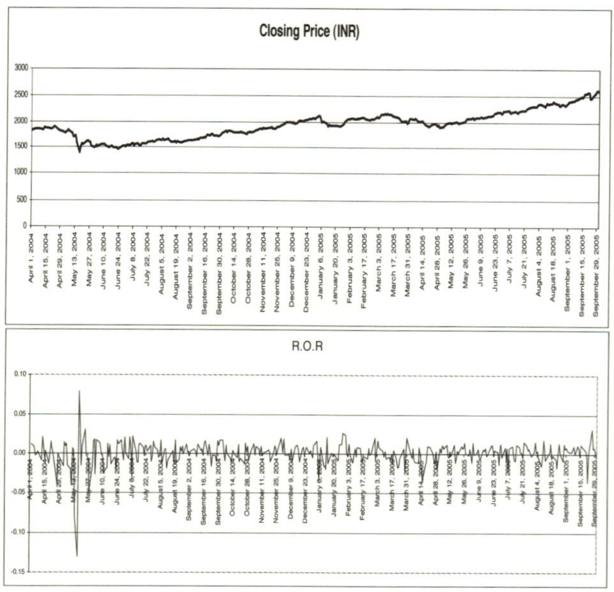

Table 1. NIFTY (April 01, 2004 - September 30, 2005)

Correlation for Price Movements		Correlation for Movements in ROR	
1-day Correlation	2-day Correlation	1-day Correlation	2-day Correlation
0.07	-0.07	0.07	-0.07

Total number of days of observation: 380
Number of days of positive ROR: 218
Number of days of negative ROR: 161

5 Observations

For Nifty the daily closing prices have been considered and for Nifty Futures the daily settlement prices have been taken into account. The correlation for the Nifty for a single time gap (1 day) as well as for two time gaps (2 days) turn out to be insignificant. For the sixteen Futures contracts also, the daily time correlations are not significant. In other words, both the cash and futures markets are on the whole efficient in the short term. This is contrary to the findings of some earlier authors and may suggest that the Indian markets have matured and gained in efficiency in recent times [5, 6]. Further, the betas for the futures contracts, representing correlation between their price movements and those of the Nifty, are close to unity. Thus, the Nifty futures contracts move more or less in tandem with the Nifty and can act as effective hedge for risk management.

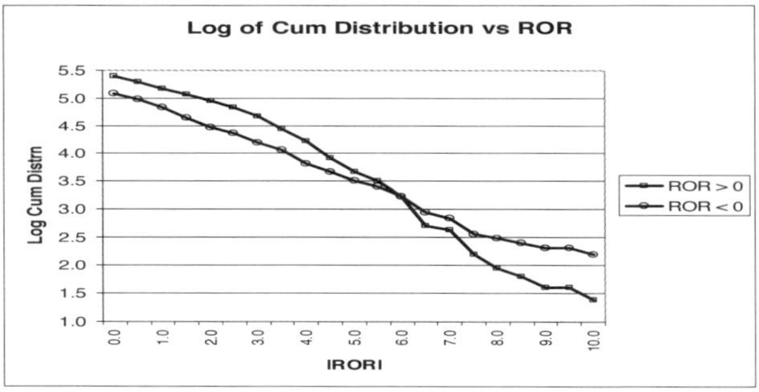

However, the basis exhibits a non-negligible correlation in most cases. There is a kind of zigzag movement of the basis over its lifetime. This is possible because the basis may decrease while both the spot and futures prices increase and vice versa. Moreover, the graphs reveal that the basis does not necessarily oscillate between positive and negative values while moving towards its maturity value of zero (as in the case of the bar attendance problem of El Farol). The basis is positive for long periods just as it is negative for long tenures. Such situations should normally open opportunities for arbitrage or reverse arbitrage between the Cash Market and Futures Market that would bring back the basis to its cost of carry. However, this apparently did not happen perhaps because such arbitrages were considered inconvenient and cumbersome. This may be explainable in part by the lack of facility for short selling in the cash market. The Security Exchange Board of India (SEBI), the market regulator, has since permitted various institutions including Foreign Institutional Investors (FII) and Mutual Funds (MF) to sell stocks without owning them. However, this alone probably cannot explain the rather large correlations between the markets. Such non-negligible correlations along with absence of arbitrage points towards the lack of adequate liquidity and market efficiency for such inter market transactions. In other words, while the cash and futures markets exhibit short-term efficiency and liquidity, inter market transactions lack such efficiency and liquidity.

Table 2. Summary of Findings

Contract	Correlation for Price Movements	Correlation for Movements of Basis	Price Beta	ROR Beta
June-04	-0.02	-0.26	1.04	1.12
July-04	0.02	-0.27	1.06	1.15
August-04	-0.08	-0.17	1.09	0.76
September-04	-0.03	-0.29	0.98	0.98
October-04	0.10	-0.23	1.02	0.92
November-04	-0.02	-0.38	1.01	1.00
December-04	0.03	-0.10	0.99	0.99
January-05	0.05	-0.29	0.96	0.97
February-05	-0.13	-0.20	1.01	1.00
March-05	-0.07	-0.40	0.96	0.96
April-05	0.02	-0.11	1.01	1.05
May-05	0.02	0.11	1.04	0.97
June-05	0.08	-0.27	1.00	1.04
July-05	0.03	-0.28	1.05	0.97
August-05	-0.06	-0.42	1.10	1.00
September-05	0.13	-0.40	1.04	1.03
October-05	0.07	-0.41	1.05	1.03
November-05	0.30	-0.04	1.06	1.02

References

1. Sornette D (2002) Why Stock Markets Crash? Princeton University Press
2. Wang Y, Wu J, Di Z (2004) Physics of Econophysics
3. Mandelbrot BB, Hudson RL (2005) The (Mis) Behaviour of Markets. Viva Books Pvt. Ltd.
4. Derivatives Market Module of the National Stock Exchange
5. Agarwal RN. Financial Integration and Capital Markets in Developing Countries A Study of Growth, Volatility and Efficiency in the Indian Capital Market
6. Pandey A. Efficiency of Indian Stock Market
7. Cooray A. Do the Stock Markets of South Asia follow a Random Walk? School of Economics, University of Tasmania, Australia.

Movements of NIFTY, NIFTY Futures and Basis

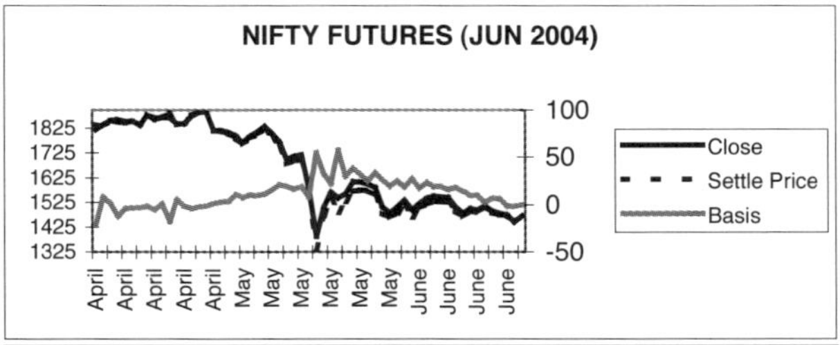

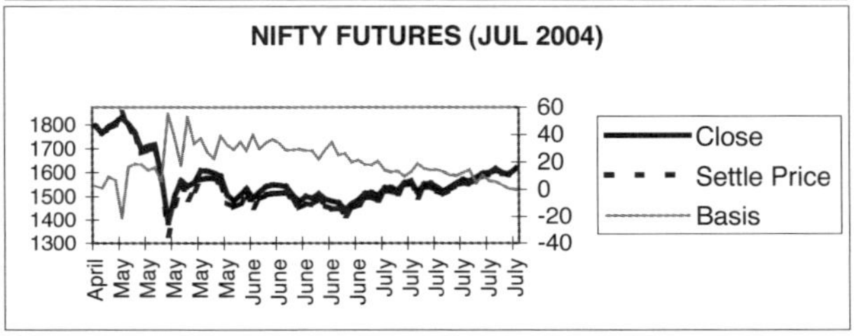

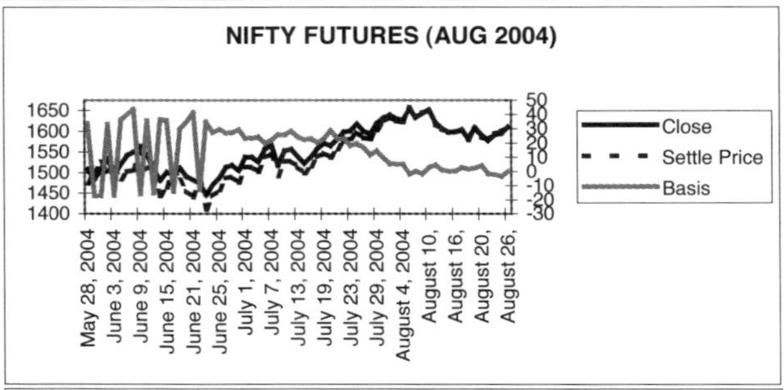

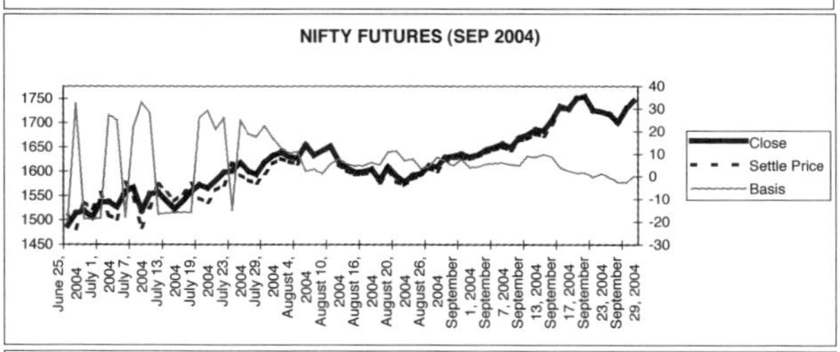

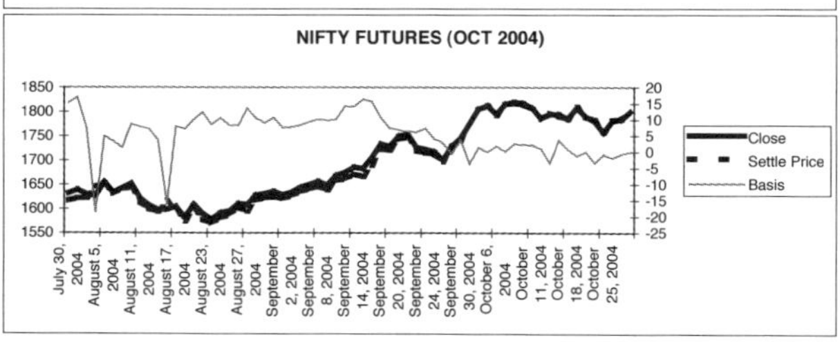

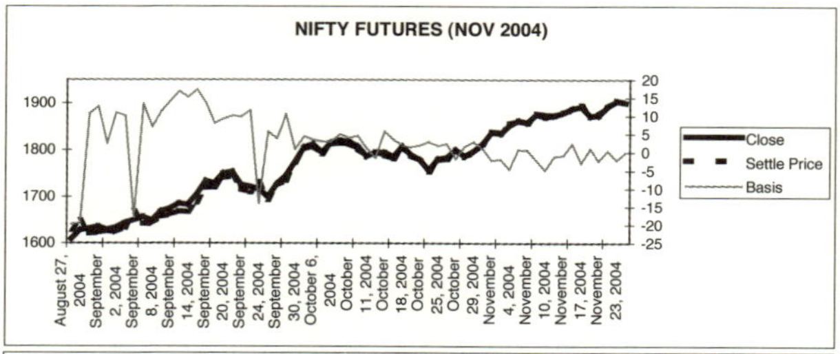
NIFTY FUTURES (NOV 2004)
Close
Settle Price
Basis

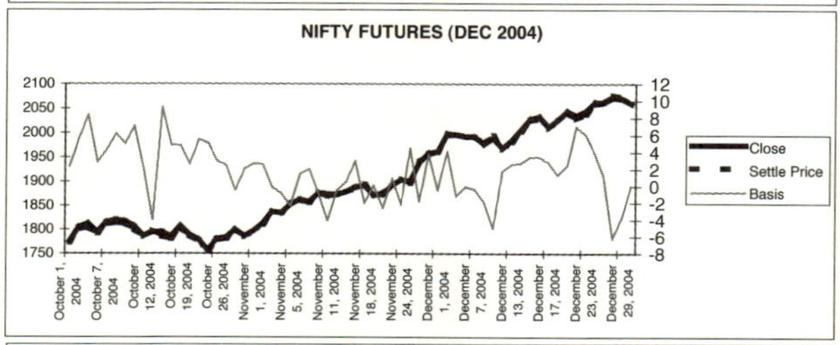
NIFTY FUTURES (DEC 2004)
Close
Settle Price
Basis

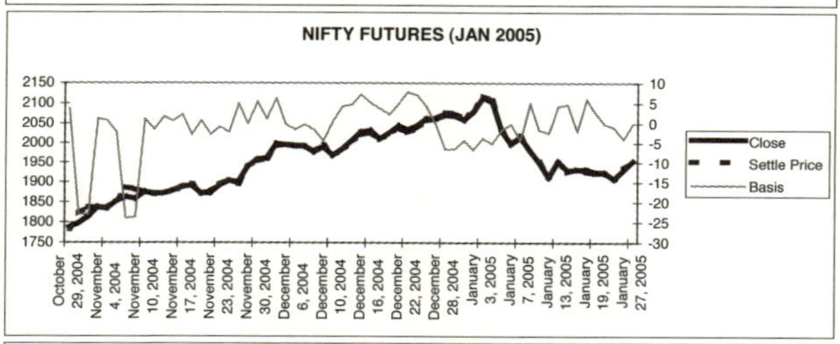
NIFTY FUTURES (JAN 2005)
Close
Settle Price
Basis

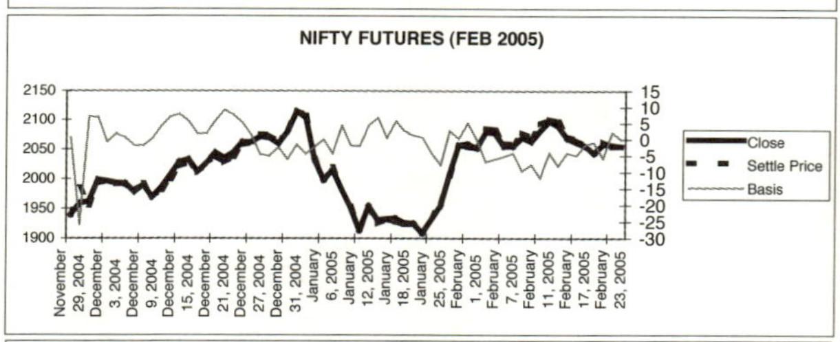
NIFTY FUTURES (FEB 2005)
Close
Settle Price
Basis

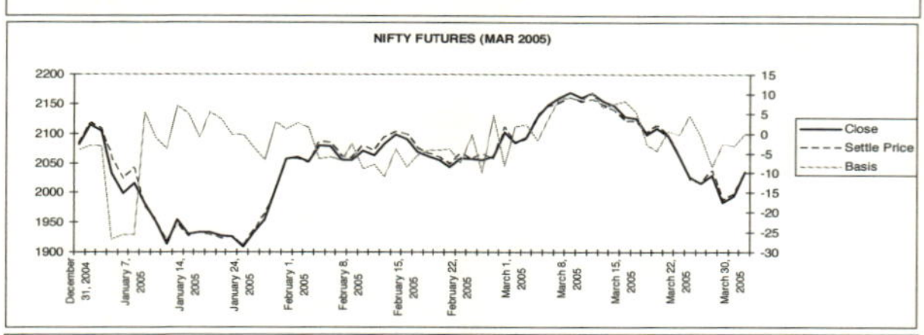
NIFTY FUTURES (MAR 2005)
Close
Settle Price
Basis

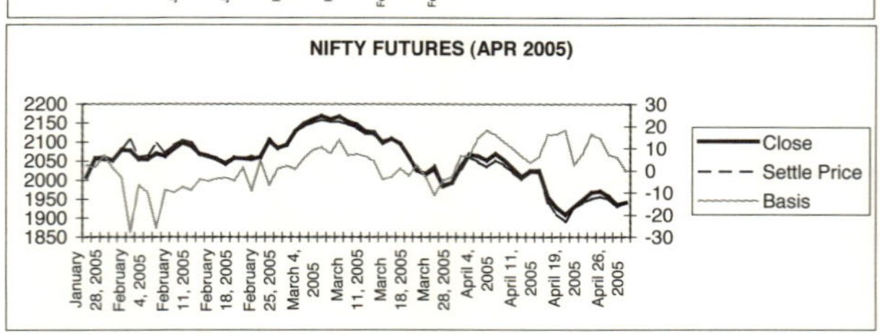
NIFTY FUTURES (APR 2005)
Close
Settle Price
Basis

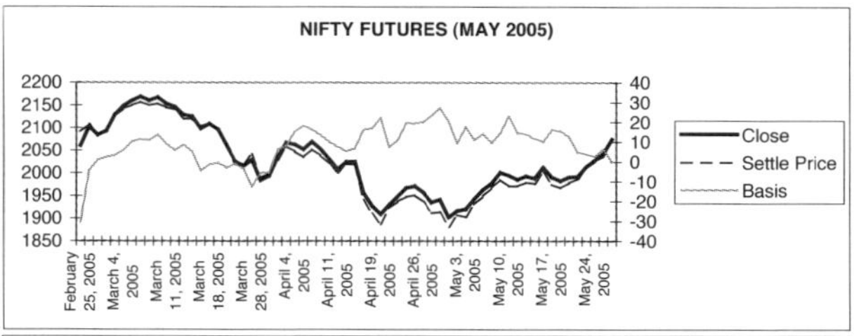
NIFTY FUTURES (MAY 2005)
2200
2150
2100
2050
2000
1950
1900
1850
40
30
20
10
0
-10
-20
-30
-40
February 25, 2005
March 4, 2005
March 11, 2005
March 18, 2005
March 28, 2005
April 4, 2005
April 11, 2005
April 19, 2005
April 26, 2005
May 3, 2005
May 10, 2005
May 17, 2005
May 24, 2005
Close
Settle Price
Basis

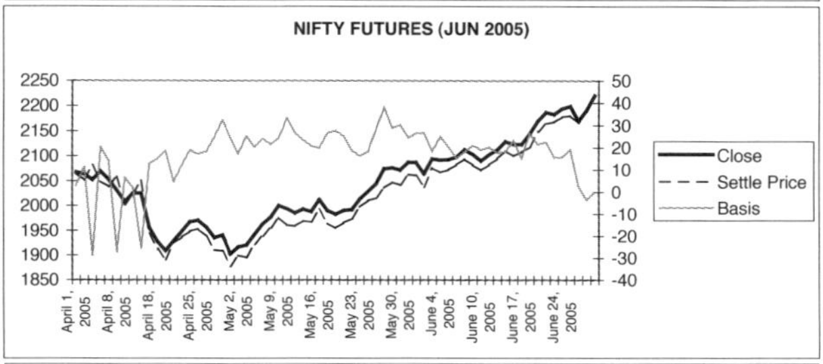
NIFTY FUTURES (JUN 2005)
2250
2200
2150
2100
2050
2000
1950
1900
1850
50
40
30
20
10
0
-10
-20
-30
-40
April 1, 2005
April 8, 2005
April 18, 2005
April 25, 2005
May 2, 2005
May 9, 2005
May 16, 2005
May 23, 2005
May 30, 2005
June 4, 2005
June 10, 2005
June 17, 2005
June 24, 2005
Close
Settle Price
Basis

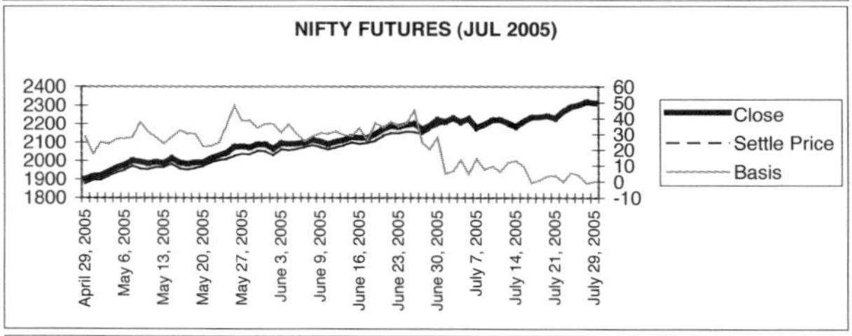
NIFTY FUTURES (JUL 2005)
2400
2300
2200
2100
2000
1900
1800
60
50
40
30
20
10
0
-10
April 29, 2005
May 6, 2005
May 13, 2005
May 20, 2005
May 27, 2005
June 3, 2005
June 9, 2005
June 16, 2005
June 23, 2005
June 30, 2005
July 7, 2005
July 14, 2005
July 21, 2005
July 29, 2005
Close
Settle Price
Basis

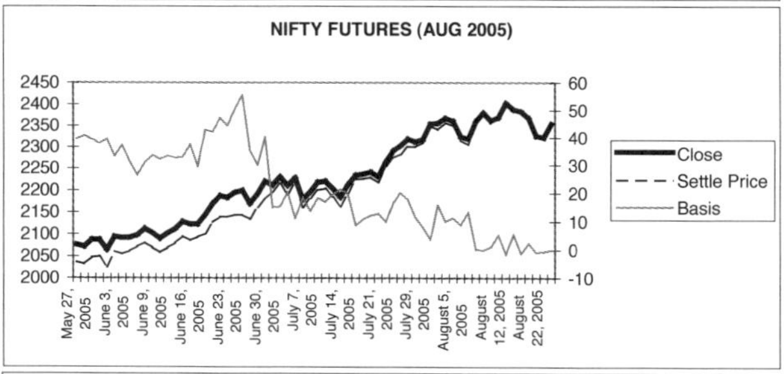
NIFTY FUTURES (AUG 2005)
2450
2400
2350
2300
2250
2200
2150
2100
2050
2000
60
50
40
30
20
10
0
-10
May 27, 2005
June 3, 2005
June 9, 2005
June 16, 2005
June 23, 2005
June 30, 2005
July 7, 2005
July 14, 2005
July 21, 2005
July 29, 2005
August 5, 2005
August 12, 2005
August 22, 2005
Close
Settle Price
Basis

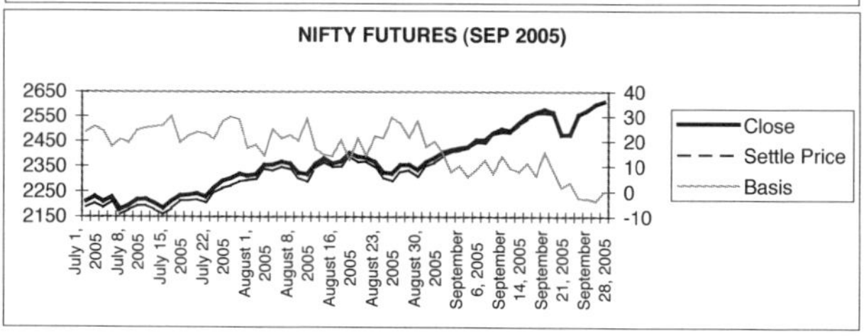
NIFTY FUTURES (SEP 2005)
2650
2550
2450
2350
2250
2150
40
30
20
10
0
-10
July 1, 2005
July 8, 2005
July 15, 2005
July 22, 2005
August 1, 2005
August 8, 2005
August 16, 2005
August 23, 2005
August 30, 2005
September 6, 2005
September 14, 2005
September 21, 2005
September 28, 2005
Close
Settle Price
Basis

Part II

Markets and their Models

Models of Financial Market Information Ecology

Damien Challet

Nomura Centre for Quantitative Finance, Mathematical Institute, Oxford University, 24–29 St Giles', Oxford OX1 3LB, United Kingdom
challet@maths.ox.ac.uk

Summary. I discuss a new simple framework that allows a more realistic modelling of speculation. The resulting model features expliciting position holding, contagion between predictability patterns, allows for an explicit measure of market inefficiency and substantiates the use of the minority game to study information ecology in financial markets.

Heterogeneous agent-based models (HAM) are inherently harder to solve analytically than traditional representative agent models (see however [1] for a review of the Economics and Behavioural Finance literature on tractable HAM). Parameter heterogeneity, for instance in the case of price return extrapolation of trend-followers, can be solved by standard methods (see [2] and variants). But for more complex heterogeneity, as when the pair-wise interaction strength varies amongst the agents, more powerful methods are needed. This is why the solution to the El Farol bar problem [3] could not be found until sophisticated methods first designed for disordered physical systems [4] were used [5, 6]; in principle, these methods solve exactly any HAM where the interaction strength is independent from the physical distance between two agents. This is why HAM are a natural meeting point of Economics and Physics. An important contribution of Econophysics to HAM is the derivation of the Minority Game (MG) [7] from the El Farol bar problem, and its numerous variants [8]. Players in a minority game are rewarded when they act against the crowd. One key ingredient of the MG is a simple, well-defined strategy space of finite dimension, made up of look-up tables prescribing a binary action for all possible market states.

Most of HAM literature, both from Economics and Physics, rests on the assumption that the agents have an action horizon of not more than one time step. In other words, they take a decision at a given time step in order to optimise some quantity at next time step. This might be justified if the considered time step is long enough. However it is obvious that real-life speculation implies that the traders have partially overlapping asset holding periods, with no systematic synchronisation. The closest approach in Economics literature is that of overlapping generation models and their numerous extensions [9–11]. The very fact that after many years almost all agent-based market models are still built with one-time step strategies is an indication that finding a strategy space of finite dimension suitable for the proper modelling of speculation is difficult. The two notable exceptions are i) a model where the agents try to predict two time-steps ahead, but cannot hold their positions [12] ii) a discussion of actual multi-time steps trading strategies (such as trend-following and value-based strategies) and more relevant to the present work, an analytical description of the evolution of the predictability associated to exploiting an isolated pattern, and on the contagion of predictability to neighbouring patterns [13]; this discussion does not include the overlapping investments.

As a consequence, the challenge is to find a strategy space that allows for inter-temporal actions, hence, that should give a good reason to the traders for opening and closing their positions, and an equally good reason for holding them. Its parametrisation should be as simple as possible so as to minimise the number of parameters, while having a dimensionality under control.

Finding a simple though meaningful speculation strategy space is surprisingly easy when examining the reasons why real traders open and close a position. It is reasonable to assume that they base their decisions on signals or patterns, such as mispricing (over/undervaluation), technical analysis, crossing of moving averages, news, etc. How to close a position is a matter of more variations: one can assume a fixed-time horizon, stop gains, stop losses, etc. I assume that the traders are inactive unless they receive some signal because some known pattern arises; this is to be compared with the many models where the agents trade at each time step [1, 7, 14]. Therefore, the agents hold their positions between patterns. All the kinds of information regarded as possibly relevant by all the traders form the ensemble of the patterns which is assumed to be countable and finite.

Each trader recognises only a few patterns, because he has access to or can analyse only a limited number of information sources, or because he does not regard the other ones as relevant; in the present model, a trader is assigned at random a small set of personal patterns which is kept fixed during the whole duration of the simulation. Every time one of his patterns arises, he decides whether to open/close a position according to his measure of the average return between all the pairs of patterns that he is able to recognise. This is precisely how people using crossings of moving averages behave: take the case of a trader comparing two exponential moving averages (EMA) EMA100 and EMA200, with memory of 100, respectively 200 days: such trader is inactive unless EMA100 and EMA200 cross; the two types of crossing define two signals, or patterns. For instance, a set of patterns could be the 20 possible crossings between EMAs with memory length of 10, 20, 50, 100 and 200 days.

The hope and the role of a trader are to identify pairs of patterns such that the average price return between two patterns is larger than some benchmark, for instance a risk-free rate (neglecting risk for the sake of simplicity); in this sense the agents follow the past average trend *between two patterns*, which makes sense if the average return is significantly different from the risk-free rate. In contrast with many other models (e.g. [14]), the agents do not switch between trend-followers/fundamentalist/noise traders during the course of the simulation.

Defining what 'average return' precisely means leads to the well-known problem of measuring returns of trading strategies in a back-testing, i.e., without actually using them. This is due to market impact and results usually in worse-than-expected gains when a strategy is used. Estimating correctly one's market impact is therefore a crucial but impossible aspect of back-testing because of two types of delusions. The first, temporal inaccuracy, is due to the over/underestimation of reaction time. Self-impact on the other hand is the impact that a real trade has on the price and market, which is not present in data used for back-testing. Both of them cause imprecision in estimating returns of strategies not being used and, accordingly, both are actively investigated by practitioners. In the following one will call naive the agents who cannot measure exactly the market impact, while sophisticated agents will be those who can.

1 Definition

The mathematical definition of the model is as follows: N traders can submit buy orders $(+1)$ or sell orders (-1) or just be inactive. They base their decisions on patterns, denoted by μ, and taking values $1, \cdots, P$. Each trader $i = 1, \cdots, N$ is able to recognise S patterns $\mu_{i,1}, \cdots, \mu_{i,S}$, drawn at random and uniformly from $\{1, \cdots, P\}$ before the simulations commence; he is active, i.e., may wish to buy or sell one share of stock, only when the current pattern $\mu(t)$ is in his pattern list, that is, $\mu(t) \in \{\mu_{i,1}, \cdots, \mu_{i,S}\}$. The kind of position he might take ($a_i(t) \in \{0, \pm 1\}$) is determined by the average price return between two consecutive occurrences of patterns. The time unit is that of pattern change, i.e., at each time step, $\mu(t)$ changes and is unique; it can depend on previous price, as in the minority game; for the time being, for the sake of simplicity, $\mu(t)$ is drawn at random and uniformly from $1, \cdots, P$.[1] The duration of a time step is assumed to be larger than the time needed to place an order. The order in which agents' actions arrive is disregarded. Therefore, at time t, the volume is $V(t) = \sum_i |a_i(t)|$ and the excess return $A(t) = \sum_{i=1}^{N} a_i(t)$ results in a (log-) price change of

$$p(t+1) = p(t) + A(t). \tag{1}$$

$p(t+1)$, not $p(t)$, is the price actually obtained by the people who placed their order at time t. Note that if the order of agents' actions were taken into account, one would first assign a random rank to the orders (possibly

[1] The history of the exact solution of minority game showed the advantage of considering Markovian μs [6, 24, 25].

influenced by agents' reaction time); then the price obtained by the n-th order would be $p(t) + A(t)n(t)/V(t)$; if the reaction of all the agents is the same, the average effect of order arrival is irrelevant on average.

There are several ways to compute returns between two patterns. Assume that $\mu(t) = \mu$ and that $t' > t$ is the first subsequent occurrence of pattern ν: $p(t') - p(t)$ is the price difference between these two patterns that does not take into account price impact, whereas $p(t' + 1) - p(t + 1)$ is the realistic price difference. Agent i perceives a cumulative price return $U_{i,\mu\to\nu}$ between pattern μ and ν which evolves according to

$$U_{i,\mu\to\nu}(t'+1) = U_{i,\mu\to\nu}(t) + p(t'+1) - p(t+1) \\ -(1-|a_i(t)|)\zeta_i[A(t') - A(t)]. \tag{2}$$

when pattern μ has appeared but not ν yet, and stays constant otherwise; ζ is the naivety factor: agents have $\zeta_i = 1$ and fail to take reaction time into account properly, while or sophisticated, agents have $\zeta_i = 0$ and compute perfectly the average price return. Finally, the variable $|a_i(t)|$ states whether the agent has opened/closed a position at time t ($|a_i(t)| = 1$), or was inactive ($|a_i(t)| = 0$). When an agent trades, he perceives perfectly the effect of his reaction time whatever ζ_i. In practice, $\zeta_i \neq 0$, and can be of any sign and value. This is because estimating reaction time exactly is impossible: even if real traders are often acutely aware of its importance, they always over or underestimate it.

An agent only opens position between two of his patterns μ and ν if the average price return between them is larger than $\epsilon > 0$, that is, if $|U_{i,\mu\to\nu}| > \epsilon t_{\mu\to\nu}$ where $t_{\mu\to\nu}$ is the total number of time-steps elapsed between patterns μ and ν.

A further specification is that an agent trades between his E best pairs of patterns, where $E \leq S(S-1)/2$ is his maximal exposure, as one cannot trade from μ to ν and from ν to μ at the same time. If $E < S(S-1)/2$, the agent aims at trading only between his most profitable patterns; in this sense, the present model could be called the *pattern game*. In the following, the simplest case $S = 2$, $E = 1$ is analysed. The dynamics becomes simpler: when $\mu(t) = \mu_{i,1}$, if $|U_{\mu_{i,1}\to\mu_{i,2}}| > \epsilon t_{\mu_{i,1}\to\mu_{i,2}}$, he buys one share ($U_{i,\mu_1\to\mu_2}(t) > 0$) or short-sells one share ($U_{i,\mu_1\to\mu_2}(t) < 0$)[2], and holds his position until $\mu(t) = \mu_{i,2}$. When an agent closes and opens the same kind of position, he simply keeps his position open.

Thus the basic model has five parameters: N, S, P, ϵ and E and has the following ingredients: the agents are adaptive and heterogeneous, they have limited cognition abilities, and they can be naive or sophisticated.

Relevant quantities include the variance of A

$$\sigma^2 = \frac{\langle A^2\rangle - \langle A\rangle^2}{P}, \tag{3}$$

---

[2] Short-selling consists in selling shares that one does not own. Closing such a position consists in buying back the stock later at a hopefully lower price.

the predictability seen by the naive agents

$$J = \frac{1}{P(P-1)} \sum_{\mu,\nu,\mu\neq\nu} \langle A(t) | \mu \to \nu \rangle^2, \tag{4}$$

where $\langle A(t)|\mu \to \nu\rangle$ stands for the average price return per time step between the occurrence of μ at time t and the next occurrence of ν. J measures predictability that the naive agents with $\zeta = 0$ hope to exploit; it is in fact related to absolute value of the average increase of $U_{\mu\to\nu}$ of naive agents per time step. Another closely related quantity is

$$K = \frac{1}{P(P-1)} \sum_{\mu,\nu,\mu\neq\nu} \langle A(t+1) | \mu \to \nu \rangle^2 \tag{5}$$

is the actual exploitable predictability which is relevant to the sophisticated agents. Finally, a measure of price bias conditional to market states is given by the average return conditional to a given pattern

$$H = \frac{\sum_\mu \langle A|\mu\rangle^2}{P}. \tag{6}$$

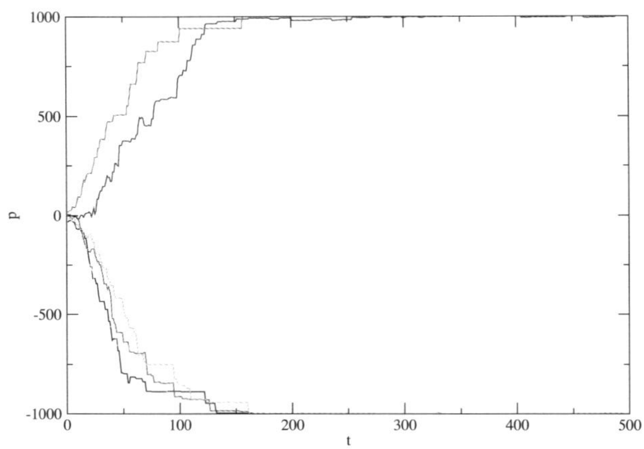

Fig. 1. Price time series with $\epsilon = 0$ ($N = 1000$, $P = 32$, $S = 2$, $\zeta = 0$)

Before carrying out numerical simulations, one should keep in mind that the price is bounded between $-N$ and $+N$, since the traders are not allowed to have an exposure larger than 1. Assume that $\epsilon = 0$, and that all the scores have small random initial valuation (otherwise nobody trades in the first place). One observes in such case beginnings of bubbles or anti-bubbles, the price rising or decreasing to $\pm N$, and then staying constant. Indeed, the price increase/decrease is echoed in the scores of all the agents, which have all the same sign, therefore all stipulate the same action. The price is stuck (Fig 1), as nobody wants to close his position, because everybody is convinced that the price should carry on on its way up/down. This phenomenon is found for all values of ζ.

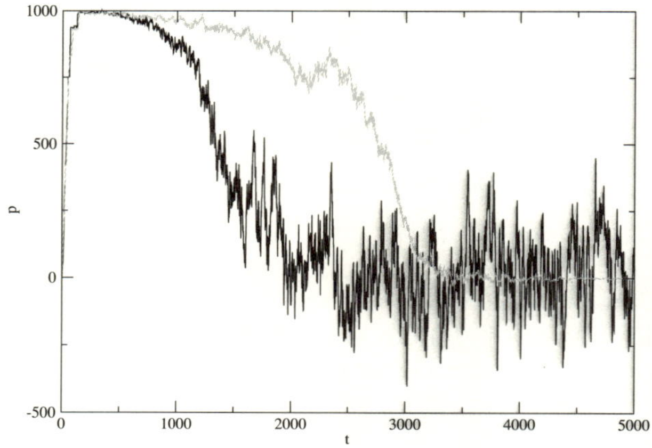

Fig. 2. Price time series with $\epsilon = 0.01$ ($N = 1000$, $P = 32$, $S = 2$). Black lines are for naive traders and green lines are for sophisticated traders.

Figure 2 illustrates the typical price time series for $\epsilon > 0$: first a bubble, then a waiting time until some traders begin to withdraw from the market. The price goes back to 0 and then fluctuates for a while. How these fluctuations are interpreted by the agents depends on ζ and ϵ: increasing ζ makes it more difficult for the agents to understand that they should refrain from trading, because they are not capable of disentangling their own contribution from these fluctuations. Accordingly, the larger ζ, the later the agents withdraw from the market, and the smaller ϵ, the longer it takes (Fig. 3). In this figure, the maximum number of iteration was capped at 10^6; naive agents need therefore a very long time before withdrawing if ϵ is small. The scaling $T_w \propto N/\epsilon$ holds only for small ζ and ϵ. For large ϵ, ζ does not matter much.

All the agents eventually withdraw from the market. This makes complete sense, as there is no reason to trade. Naivety results in a diminished ability to withdraw rapidly enough from a non-predictable market, and, as a by-product, in larger price fluctuations. This is consistent with the fact that volatility in real markets is much larger than if the traders were as rational as mainstream Economics theory assumes (see e.g. [15]). Naivety, an unavoidable deviation from rationality, is suggested here as one possible cause of excess volatility.

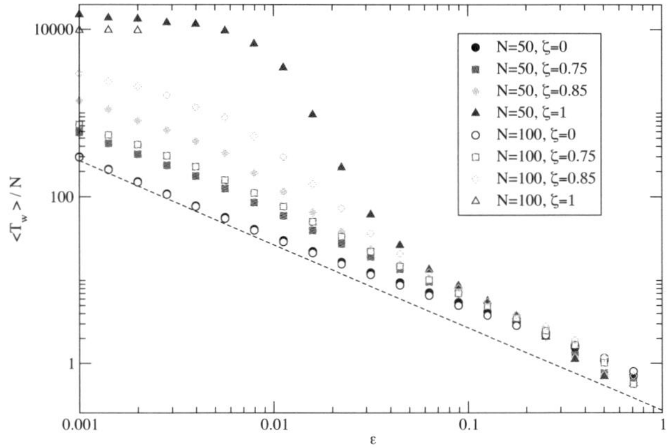

Fig. 3. Scaled time to withdraw as a function of ϵ. $P = 10$, $N = 50$ (full symbols) and 100 (empty symbols), $\zeta = 0$ (circles), 0.75 (squares), 0.85 (diamonds) and 1 (triangles); average over 500 samples

2 Noise traders

As the traders try and measure average returns, adding noise to the price evolution ($A(t) \to A(t) + N_n\eta(t)$) where $\eta(t)$ is uniformly distributed between -1 and 1 does not provide any trading opportunity, but makes it more difficult to estimate precisely average returns. Accordingly, the agents withdraw later, and the larger ζ, the later. This is an additional clue that naive agents are blinded by the fluctuations that they produce themselves.

3 Market impact heterogeneity

Real traders are heterogeneous in more than one way. In a population where each trader has his own ζ_i, people with a small ζ_i evaluate gain opportunities better. This also means that the impact of people with large ζ_i provides predictability to the agents with a lower ζ_i, and therefore the former are exploited by the latter, giving a good reason to trade to sophisticated agents as long as naive agents are active.

4 Market information structure

Up to now, the model showed how hard it is not to trade. But how hard is it to make the agents want to trade? The problem is that they try to detect and exploit predictability, but that there is none when all the agents have the same abilities (e.g. the same ζ_i). They might be fooled for a while by price fluctuations, as they try to detect trends between patterns, but eventually realise that there is no reason to trade. This is why the heterogeneity of the agents is crucial in this model. For instance, sophisticated agents would continue to trade if new naive agents replaced those who had understood their mistake. This would however probably preclude any hope to solve the model exactly. This is why I shall assume a different kind of heterogeneity. I will assume that there are people, the producers, who need to trade for other reasons than mere speculation [16, 17]: they use the market but make their living outside of it. Doing so they do not withdraw from the market, inject predictability and are exploited by the speculators. A simple way to include N_p producers in the model is to assume that they induce a bias in the excess demand that depends on $\mu(t)$, i.e., $A(t) = A_{\text{prod}}^{\mu(t)} + A_{\text{spec}}(t)$. Each producer has a fixed contribution ± 1 to A_{prod}^{μ}, drawn at random and equiprobably from $\{-1, +1\}$ for each μ and each producer, hence $A_{\text{prod}}^{\mu} \propto \sqrt{N_p}$. In that way, the amount of market impact predictability introduced in the model is well controlled.

If there is no information structure in the market, i.e., if $\mu(t)$ does not depend at all on past patterns or price history, the effect of producers is akin to that of noise traders, hence, the speculators cannot exploit the predictability left by the producers. This is because the speculators need temporal correlations between occurring biases in order to exploit them, which happens when the transitions between market states are not equiprobable, i.e., when the transition matrix between patterns W is such that $W_{\mu \to \nu} \neq 1/P$. This assumption is supported by empirical results: [18] determined states of the market with a clustering method, and found that the transitions between the states is highly non-random and has long-term memory which one neglects. Numerically, one chose to fix $W_{\mu \to \nu}$ to a random number between 0 and 1 and then normalised the transition probabilities; the variance of $W_{\mu \to \nu}$ is a parameter of the model which controls the amount of correlation between biases induced by the producers. It should be noted that the speculators are not able remove the global price bias because of the linearity of the price impact function. As a consequence, one should ensure that the effective average price return bias introduced by the producers is zero; it is nothing else but the sum of the components of A_{prod} weighted by the frequency of appearance of μ, which is obtained from W^{∞}.

Adding more producers, that is, more predictable market impact, increases linearly the values of inter-pattern price return predictability measures J and K, as well as the average price impact H and the fluctuations of the price σ^2. Then, keeping fixed the number of producers and their actions, adding more and more sophisticated speculators first decreases the fluctuations σ^2, the price impact H, and the exploitable predictability measures J and K (Fig. 4); then all these quantities reach a minimum at an N of order P^2, that is, of the order of the number of pairs of patterns. At the same time, the average total *number* of speculators in the market, denoted by $\langle N_{\text{act}} \rangle$, reaches a plateau, thus the speculators are able to refrain from trading when predictability falls below some value determined by ϵ. The fact that the fluctuations σ^2 increase for $N > P(P-1)/2$ means that the agents with the same pair of patterns enter and withdraw from the market in a synchronous way, leaving $\langle N_{\text{act}} \rangle$, H, J and K constant. Avoiding such synchronicity can be realised by letting the agents have heterogeneous ϵ_i or ζ_i. The average gain of the speculators is positive when their number is small enough but becomes negative for at $N \simeq P(P-1)/2$ in Fig 4. All these properties are found in grand canonical minority games [19]. If evolving capitals were implemented, less speculators would survive, which would lower the effective N, thus keeping the average gain at or above zero. This would also reduce or suppress the increase of σ^2.

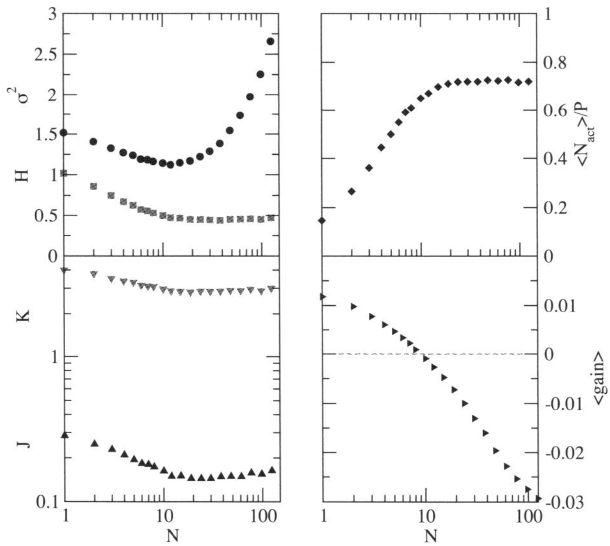

Fig. 4. Volatility σ^2 (circles), price impact H/P (squares), naive predictability J (down triangles) and sophisticated predictability K (up triangles), scaled fraction of active speculators (diamonds), and average gain per speculator and per time step (right triangles); $P = 5$, $N_p = 10$, $\epsilon = 0.05$; average over 10000 samples of speculators

The relationship between the producers and the speculators can be described as a symbiosis: without producers, the speculators do not trade; without speculators, the producers lose more on average, as the speculators reduce H. This picture, remarkably similar to that of the Minority Game with producers [17,20,21], justifies fully in retrospect the study of information ecology with minority games.

Guided by the knowledge that any mechanism subordinated to a random walk and responsible for making the agents switch between being in and out of the market produces volatility clustering [22], one expects the emergence of such property in this model when the number of agents is large and $\epsilon > 0$. This is exactly what happens, as shown by Fig. 5, where the volume $N_{\rm act}$ displays a long term memory. Whether this model is able to reproduce faithfully real-market phenomenology is under investigation.

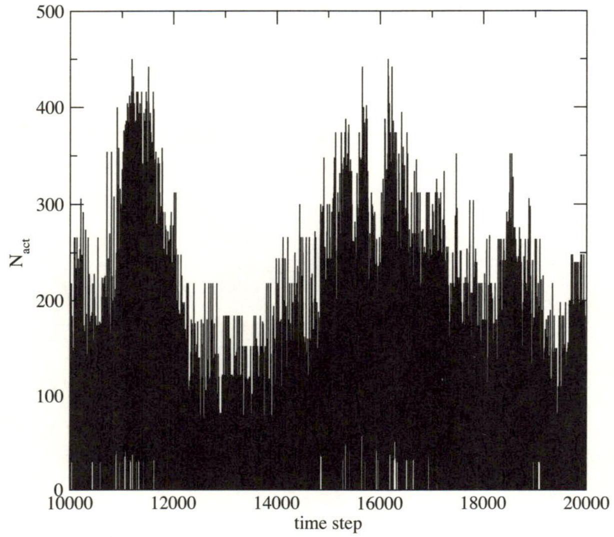

Fig. 5. Number of active speculators as a function of time ($P = 10$, $N = 1000$, $N_p = 100000$, $\epsilon = 0.001$)

The similarity of information ecology between this model and the MG is a clue that the MG has some relevance for financial market modelling, and suggests to reinterpret it. Generally speaking, a minority mechanism is found when agents compete for a limited resource, that is, when they try to determine by trial and error the (sometimes implicit) resource level, and synchronise their actions so that the demand equals the resource on average [5, 23]. As a consequence, an agent is tempted to exploit the resource more than his fair share, hoping that the other agents happen to take a smaller share of it. The limited resource in financial markets is *predictability* and indeed information ecology has proven to be one of the most fascinating and plausible insights of minority games into market dynamics [17, 20]. Instead of regarding $A(t)$ in the MG as the instantaneous excess demand, one should reinterpret it as the *deviation from unpredictability* $A = 0$ at time t. The two actions can be for instance to exploit an inefficiency (+1) or to refrain from it (−1); A in this context would measure how efficiently an inefficiency is being exploited. Then everything becomes clearer: the fact that $A(t)$ display mean-reverting behaviour is not problematic any more as it is not a price return. It simply means when the agents tend to under-exploit some predictability, they are likely to exploit it more in the next time steps, and reversely. What price return correspond to a given A is not specified by the MG, but herding in the information space (the MG) should translate into interesting dynamics of relevant financial quantities such as volume and price; for instance, dynamical correlations of $|A|$ in the MG probably correspond to dynamical correlations in the volume of transactions. Therefore, studying the building up of volatility feedback, long-range memory and fat-tailed A still makes sense, but not in a view to model explicitly price returns.

5 Conclusion

This model provides a new simple yet remarkably rich market modelling framework. It is readily extendable, and many relevant modifications are to be studied. The similarity of information ecology between the present model and the MG is striking and allowed for reinterpretation of the MG as model of competition for predictability, and substantiated the use of the MG to study predictability ecology of financial markets. Whether the proposed model is exactly solvable is under scrutiny.

References

1. Hommes CH (2005) in Handbook of Computational Economics, Ed. Judd KL, Tesfatsion L, Elsevier Science B. V.
2. Brock WA, Hommes CH (1997) Econometrica 65:1059
3. Arthur WB (1994) Am. Econ. Rev. 84:406
4. Mézard M, Parisi G, Virasoro MA (1987) Spin glass theory and beyond, World Scientific
5. Challet D, Ottino G, Marsili M (2004) Physica A 332:469; preprint cond-mat/0306445
6. Coolen ACC (2005) The Mathematical Theory of Minority Games, Oxford University Press, Oxford
7. Challet D, Zhang Y-C (1997) Physica A 246:407; adap-org/9708006
8. Challet D (2005)
9. Allais M (1947) Économie et intérêt, Imprimerie Nationale, Paris
10. Samuelson P (1958) J. of Political Economy 66:467
11. Geneakoplos J, Polemarchakis P (1991) in The Handbook of Mathematical Economics, Ed. Hildebrand W, Sonnenschein H, North Holland, London, IV:1899–1962
12. Rothenstein R, Pawelzik K (2003) Physica A 326:534
13. Farmer JD (1999) Tech. Rep. No. 98-12-117, Santa Fe Institute
14. Lux T, Marchesi M (1999) Nature 397:498
15. Shiller RJ (1981) Am. Econ. Rev. 71:421
16. Zhang Y-C (1999) Physica A 269:30
17. Challet D, Marsili M, Zhang Y-C (2000) Physica A 276:284; cond-mat/9909265
18. Marsili M (2002) Quant. Fin. 2
19. Challet D, Marsili M (2003) Phys. Rev. E 68:036132

20. Challet D, Chessa A, Marsili M, Zhang Y-C (2000) Quant. Fin. 1:168; cond-mat/0011042
21. Zhang Y-C (2001) cond-mat/0105373
22. Bouchaud J-P, Giardina I, Mézard M (2001) Quant. Fin. 1:212
23. Challet D (2004) Physica A 334:24
24. Cavagna A (1999) Phys. Rev. E 59:R3783; cond-mat/9812215
25. Challet D, Marsili M, Zhang Y-C (2005) Minority Games, Oxford University Press, Oxford.

Estimating Phenomenological Parameters in Multi-Assets Markets

Giacomo Raffaelli[1] and Matteo Marsili[2]

[1] Dipartimento di Fisica, CNR-ISC and INFM-SMC,
Universitá di Roma "La Sapienza", p.le A. Moro 2, 00185 Roma, Italy
[2] The Abdus Salam International Centre for Theoretical Physics
Strada Costiera 14, 34014 Trieste, Italy

Summary. Financial correlations exhibit a non-trivial dynamic behavior. This is reproduced by a simple phenomenological model of a multi-asset financial market, which takes into account the impact of portfolio investment on price dynamics. This captures the fact that correlations determine the optimal portfolio but are affected by investment based on it. Such a feedback on correlations gives rise to an instability when the volume of investment exceeds a critical value. Close to the critical point the model exhibits dynamical correlations very similar to those observed in real markets. We discuss how the model's parameter can be estimated in real market data with a maximum likelihood principle. This confirms the main conclusion that real markets operate close to a dynamically unstable point.

1 Introduction

The statistical analysis of financial time series has revealed fluctuation phenomena of great interest both to investors trying to exploit regularities, and to economists and physicists trying to uncover the origin of these properties. Perhaps the best known of these "stylized facts" relate to the existence of fat tails in the distribution of assets' returns, and to the temporal structure of the fluctuations (volatility clustering) [1, 2]. These features pose practical problems for risk management and option pricing. Most interestingly though, it rises the question of understanding why do returns undergo a stochastic process with these features.

Going from single asset to multi-asset markets, the relevant question becomes that of explaining the origin of financial correlations across stocks, which have been observed in empirical studies. The structure of correlations have been analyzed with several methods [3–6] and, in terms of its spectral decomposition, it is composed of three components: *1)* noise background, which accounts for the bulk of the distribution, *2)* economic correlations, which manifests in the few eigenvalues which leak out of the noise background and *3)* the largest eigenvalue Λ – corresponding to the so-called *market mode* – which is well separated from the other ones as it accounts for a significant part of the correlations. The market mode describes the co-movement of stocks and, as shown in Fig. 1a, exhibits a non-trivial dynamical behavior.

It is reasonable to assume that the properties of smaller eigenvalues are due to exogenous economic factors or to noise trading. The key question is then whether the wild fluctuations in Λ of Fig. 1a are due to exogenous factors or to the internal non-linear dynamics of the market. This issue has been addressed in Ref. [7], starting from the following considerations: One important function of financial markets is that it allows traders to "dump" risk into the market by diversifying their investment across stocks, as postulated by portfolio theory [8]. This produces a flow of investment which is correlated across assets, with a sizeable component on the direction of the optimal portfolio. This, in turn, is likely to contribute to the correlation of returns, on which portfolio optimization depends. Such a feedback on financial correlations is captured by a symbolic equation for the covariance matrix $\hat{C}$:

$$\hat{C} = \hat{\Omega} + \hat{B} + \hat{F}\left(\hat{C}\right). \qquad (1)$$

The idea is that asset correlations result from three different sources, corresponding to the three different components discussed above: *1)* noise (including speculative) trading , which accounts for $\hat{\Omega}$; *2)* fundamental trading, based on economics, represented by $\hat{B}$ and *3)* the term $\hat{F}$ due to investment in risk minimization strategies, which itself depends on the financial correlations $\hat{C}$. Eq. (1) depicts how "bare" economic correlations are "dressed" by the effect of financial trading and can be formalized in a simple phenomenological closed model for the joint evolution of returns and correlations [7]. This model predicts that, when the strength of the component of optimal portfolio investment increases, the market approaches a dynamic instability. It is precisely close to this point that the model reproduces time series of correlations with quite realistic properties (see Fig. 1b).

The picture which this model offers is that of a market where risk minimization strategies contribute to the correlations they are trying to elude. The stronger the investment activity on these strategies, the more market's correlations grow, up to a point where the market enter a dynamically unstable phase. Interestingly, close to the phase transition the model develops anomalous fluctuations also in returns, which acquire a fat tailed distribution.

Within this simplified picture, real markets happen to operate close to the critical point corresponding to the dynamic instability. This is likely not a coincidence: Indeed as far as a markets is far from the critical point, it offers stable correlations which allow traders to reliably use risk minimization strategies. However, as this opportunity is exploited more and more, the market approaches a point of infinite susceptibility where correlations become more and more unstable thus deterring further trading on these strategies.

The similarity of the model's dynamical behavior close to the phase transition with real data can be put on firmer basis by maximum likelihood estimation of the model's parameter on real market data. The rest of this paper is devoted to a detailed discussion of the fitting procedure and of its tests on synthetic data sets.

2 The model

In order to capture the considerations behind Eq. (1), let us consider a set of N assets. We imagine that the return undergo a stochastic process [9]:

$$|x(t+1)\rangle = |x(t)\rangle + |\beta(t)\rangle + \xi(t)|z(t)\rangle. \tag{2}$$

the term $|\beta(t)\rangle$, which is the vector of *bare* returns, i.e. it describes the exogenous factors which drive the prices, including fundamental economic processes. $|\beta(t)\rangle$ is assumed to be a Gaussian random vector with

$$E[|\beta(t)\rangle] = |b\rangle, \qquad E[|\beta(t)\rangle\langle\beta(t')|] = |b\rangle\langle b| + B\hat{I}\delta_{t,t'} \tag{3}$$

where $\hat{I}$ is the identity matrix and $|b\rangle$ and B will be considered as parameters in what follows.

The other term captures inter-asset correlations emerging from portfolio investments. $\xi(t)$ is an independent Gaussian variable with mean fixed ϵ and variance Δ. The vector $|z(t)\rangle$, which describes the direction along which portfolio investment affects the inter-asset correlations, is the optimal portfolio calculated with standard portfolio optimization techniques. So, the vector $|z(t)\rangle$ will be determined by a risk minimization procedure for fixed expected return R (expressed in money units, not in percentage, so that $|z(t)\rangle$ itself has the meaning of money invested in each asset, not of percentage of one investor's wealth). We also fix the total wealth $\langle z|1\rangle = W$ invested, and keep it fixed so that $|z\rangle$ will be the solution of

$$\min_{|z\rangle,\nu,\sigma} \left[\frac{1}{2}\langle z|\hat{C}(t)|z\rangle - \nu\left(\langle z|r(t)\rangle - R\right) - \sigma\left(\langle z|1\rangle - W\right)\right] \tag{4}$$

where $\hat{C}(t)$ is the correlation matrix at time t. Both the expected returns and the correlation matrix are computed from historical data over a characteristic time τ:

$$|r(t+1)\rangle = (1-e^{-1/\tau})|r(t)\rangle + e^{-1/\tau}|\delta x(t)\rangle \tag{5}$$
$$\hat{C}_1(t+1) = (1-e^{-1/\tau})\hat{C}_1(t) + e^{-1/\tau}|\delta x(t)\rangle\langle\delta x(t)| \tag{6}$$
$$\hat{C}(t+1) = \hat{C}_1(t+1) - |r(t)\rangle\langle r(t)| \tag{7}$$

where $|\delta x(t)\rangle \equiv |x(t+1)\rangle - |x(t)\rangle$. This makes the set of equations above a self-contained dynamical stochastic system.

This model can be solved in the $\tau \to \infty$ limit [7]. In this case, the model has a singular behavior for Λ when the volume of investment W reaches a critical threshold

$$W_c = \frac{N}{2\epsilon}\left[\sqrt{\frac{\langle b|b\rangle + 4\epsilon R}{N}} - \frac{\langle b|1\rangle}{N}\right]. \tag{8}$$

The singularity manifests itself with the emergence of instabilities in the dynamics of Λ. As shown in Fig. 1, the fluctuations produced by the model close to the critical point are very similar to the ones observed in real markets.

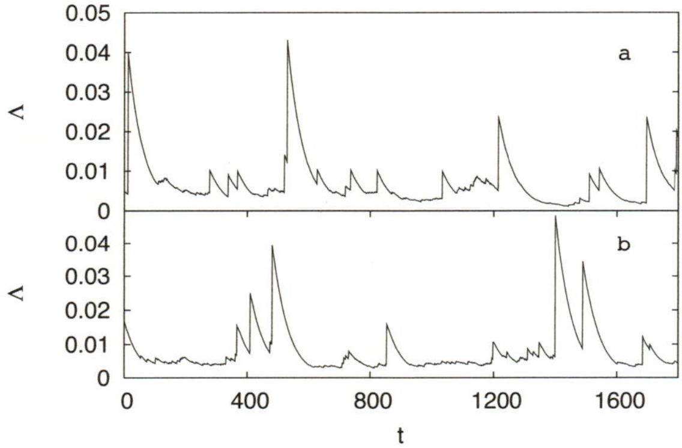

Fig. 1. Maximum eigenvalue of the correlation matrix as a function of time for $\tau = 50$. a) Toronto Stock exchange [10]. Here the correlation matrix is obtained using Eq. (6) with $|\delta x_t\rangle$ taken from historical data. b) simulation of Eq.(2) with $N = 20$, $R = 10^{-2}$, $B = 10^{-6}$,$\Delta = 0.04$,$\epsilon = 10^{-1}$, $W = 1.4$. Components of $|b\rangle$ where generated uniformly in the interval $[0, 2 \cdot 10^{-3}]$, resulting in $W^* \approx 1.41$.

3 Fitting algorithm

The similar anomalous behavior of the temporal evolution of correlations in simulations and in real markets shown in Fig. 1, suggests that markets operate close to the critical point W_c for which the data in Fig. 1b were obtained. Here we will try to put this conclusion on firmer basis by fitting real data with our model. Let us rewrite the model as

$$|\delta x(t)\rangle = |b\rangle + |\eta(t)\rangle + [1 + \zeta(t)]|y(t)\rangle \tag{9}$$

where $|\delta x(t)\rangle$ is the vector of daily returns, $|\eta\rangle$ is a zero average gaussian vector with i.i.d. components of variance B, $\zeta(t)$ is gaussian i.i.d. with zero average and variance $D = \Delta/\epsilon^2$ and $|y(t)\rangle$ is the portfolio with $\langle y(t)|r(t)\rangle = \rho = \epsilon R$ and $\langle y(t)|1\rangle = \omega = \epsilon W$. Notice that $|y(t)\rangle$ depends on ρ and ω and it is computed from the past data set's correlation matrix and returns. We compute the likelihood

$$P\{|\delta x\rangle||b\rangle, |y\rangle\} = \left\langle \prod_{i,t} \delta\left(b_i + \eta_i(t) + [1+\zeta(t)]y_i(t) - \delta x_i(t)\right) \right\rangle_{\eta,\zeta} \tag{10}$$

i.e. the probability of obtaining the observed returns $|\delta x(t)\rangle$ from the model of Eq. (9) given the parameters $|b\rangle$, $|y(t)\rangle$. In principle, $|y(t)\rangle$ depends on $|\delta x(t)\rangle$ itself through the returns $|r(t)\rangle$ and correlation matrix $\hat{C}(t)$. Here, we neglect this dependence and take $|r(t)\rangle$ and $\hat{C}(t)$ as exogenous parameters. With this assumption, $|y(t)\rangle$ becomes a function of the parameters ρ and ω (see below).

The gaussian integrals in Eq. (10) are easily carried out, and we get

$$\mathcal{L} \equiv \log P\{|\delta x\rangle||b\rangle, |y\rangle\} = -\frac{NT}{2}\log(2\pi B) + \frac{1}{2B}\sum_t F(t) \tag{11}$$

with

$$F(t) = \left\{ \frac{\langle g(t)|y(t)\rangle^2}{1/\mu + \langle y(t)|y(t)\rangle} - \langle g(t)|g(t)\rangle - B\log(1+\mu\langle y(t)|y(t)\rangle) \right\}$$

where we defined $|g(t)\rangle = |b\rangle + |y(t)\rangle - |\delta x(t)\rangle$ and $\mu \equiv D/B = \Delta/(B\epsilon^2)$. Setting to zero the partial derivative of $\mathcal{L}$ wrt b_i, we find

$$|b\rangle = \frac{1}{T}\sum_t \left[|\delta x(t)\rangle - \frac{1+\mu(\langle \delta x(t)|y(t)\rangle - \langle b|y(t)\rangle)}{1+\mu\langle y(t)|y(t)\rangle}|y(t)\rangle \right] \tag{12}$$

Setting to zero the partial derivative of $\mathcal{L}$ with respect to B yields

$$B = -\frac{2\pi}{NT}\sum_t \left(\frac{\mu\langle g|y\rangle^2}{1+\mu\langle y|y\rangle} - \langle g|g\rangle \right) \tag{13}$$

whereas μ is given by the solution of

$$\mu = \frac{\sum_t \frac{\langle g|y\rangle^2 - B\langle y|y\rangle}{(1+\mu\langle y|y\rangle)^2}}{\sum_t \frac{B\langle y|y\rangle^2}{(1+\mu\langle y|y\rangle)^2}} \tag{14}$$

With respect to the values of ρ and ω, we observe the solution $|y\rangle$ of portfolio optimization problem can be written explicitly in terms of ρ and ω as

$$|y(t)\rangle = \rho|\hat{\rho}(t)\rangle + \omega|\hat{\omega}(t)\rangle \tag{15}$$

where

$$|\hat{\rho}\rangle = \hat{C}^{-1}\frac{\chi_{11}|r\rangle - \chi_{r1}|1\rangle}{\chi_{rr}\chi_{11} - \chi_{r1}^2}, \qquad |\hat{\omega}\rangle = \hat{C}^{-1}\frac{-\chi_{r1}|r\rangle + \chi_{rr}|1\rangle}{\chi_{rr}\chi_{11} - \chi_{r1}^2} \tag{16}$$

where $\chi_{ab} \equiv \langle a|\hat{C}^{-1}|b\rangle$ and, here and in what follows, the argument t is suppressed, when not needed. The two vectors $|\hat{\rho}(t)\rangle$ and $|\hat{\omega}(t)\rangle$ can explicitly be computed directly from the data-set. This means in practice that it is enough, for each t to solve two equations of the type $\hat{C}(t)|x\rangle = |a(t)\rangle$ for all values of ρ and ω. The likelihood can be expressed explicitly in terms of ρ and ω and the partial derivatives $\partial\mathcal{L}/\partial\rho$ and $\partial\mathcal{L}/\partial\omega$ can be computed. In order to do this, it is convenient to write $|g\rangle = |h\rangle + |y\rangle$, so that $\mathcal{L}$ depends on ρ and ω only through the combinations $\langle h|y\rangle$ and $\langle y|y\rangle$ so that, e.g.

$$\frac{\partial\mathcal{L}}{\partial\rho} = \sum_t \frac{\partial\mathcal{L}}{\partial\langle h|y\rangle}\langle h|\hat{\rho}\rangle + 2\sum_t \frac{\partial\mathcal{L}}{\partial\langle y|y\rangle}\langle y|\hat{\rho}\rangle \tag{17}$$

where we used that $\partial\langle h|y\rangle/\partial\rho = \langle h|\hat{\rho}\rangle$ and $\partial\langle y|y\rangle/\partial\rho = 2\langle y|\hat{\rho}\rangle$. Defining

$$H_a(\rho,\omega) = \sum_t \frac{\partial\mathcal{L}}{\partial\langle h|y\rangle}\langle h|a\rangle, \quad L_{ab}(\rho,\omega) = \sum_t \frac{\partial\mathcal{L}}{\partial\langle y|y\rangle}\langle a|b\rangle, \quad a,b = \hat{\rho},\hat{\omega}$$

we have

$$\rho = \frac{1}{2}\frac{H_\omega L_{\hat{\rho}\hat{\omega}} - H_{\hat{\rho}} L_{\hat{\omega}\hat{\omega}}}{L_{\hat{\rho}\hat{\rho}}L_{\hat{\omega}\hat{\omega}} - L_{\hat{\rho}\hat{\omega}}^2}, \qquad \omega = \frac{1}{2}\frac{H_{\hat{\rho}} L_{\hat{\rho}\hat{\omega}} - H_{\hat{\omega}} L_{\hat{\rho}\hat{\rho}}}{L_{\hat{\rho}\hat{\rho}}L_{\hat{\omega}\hat{\omega}} - L_{\hat{\rho}\hat{\omega}}^2}$$

These are complicated non-linear equations because the left hand sides depend on ρ and ω. In particular, it is not guaranteed that the maximum is unique, so the outcome of the maximization procedure might depend on the particular procedure adopted. The algorithms which we found provides the most reliable results performs the maximization iteratively in two steps: we take, as starting values, $|b\rangle$ as the average drift of the assets in the period considered and B is taken to be the mean variance. Then, *1)* from a particular estimate of $|b\rangle$ and B we perform a maximization procedure on the function $\mathcal{L}$, for example using a simplex algorithm, with respect to the parameters μ, ω and ρ. From these, we compute a new estimate of $|b\rangle$ solving Eq. (12) as $|b\rangle = (\hat{I} - \hat{A})^{-1}|a\rangle$ with

$$A_{ij} = \frac{\mu}{T}\sum_t \frac{y_i(t)y_j(t)}{1 + \mu\langle y(t)|y(t)\rangle}$$

and

$$|a\rangle = \frac{1}{T}\sum_t \left(|x(t)\rangle - \frac{1 + \mu\langle x(t)|y(t)\rangle}{1 + \mu\langle y(t)|y(t)\rangle}|y(t)\rangle\right).$$

With these, we perform again the minimization of $\mathcal{L}$ to get new values for μ, ω and ρ, and so on until convergence. Actually, a smoothing parameter α is introduced, so that at each iteration the parameters are taken to be α times their new value plus $(1 - \alpha)$ times their old value.

Since the procedure as well as the function being minimized is quite complex, we performed some tests: We generate an artificial time series from the model itself, with known values of the parameters, and then we run the procedure to fit this data. In Fig. 2 we plot the result of such a fit. Each point corresponds to one period of $T = 100$ data points, that are fitted to obtain the parameters. The results shown are for ω, ρ and $\omega_c \equiv \epsilon W_c$, and they are quite close to the real values. In this case both the time scale used to generate the data (τ) and the one used to calculate the exponentially weighted correlation matrix (τ') were the same, $\tau = \tau' = 100$, but tests with different time scales have also been performed. The stability is quite good as long as $\tau' \approx \tau$ (that is, the two time scales do not differ by an order of magnitude).

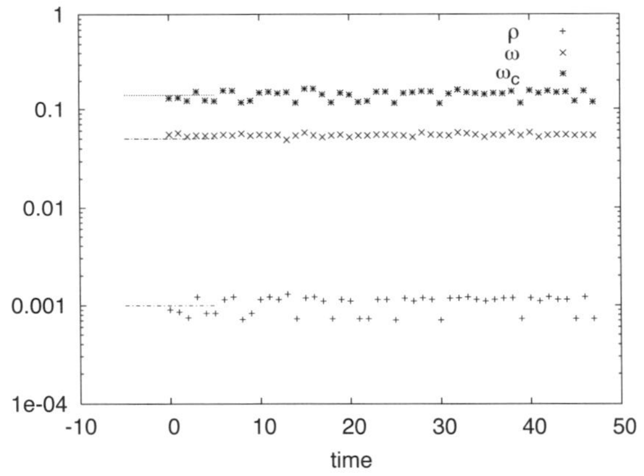

Fig. 2. Each point is a 100-days period. The data were generated with $\omega = 0.05$, $\rho = 0.001$, $\omega_c = 0.14$, $\tau = 100$, $\epsilon = 0.1$, $\Delta = 0.04$.

As a further test, in Fig.3 we plot the results of the fit for different values of ω and over a longer time window of $T = 500$ points. Here the horizontal axis reports the value of ω which was used to generate the data (ω_r), while on the vertical axis we plot the fitted value, ω_f. These results were obtained without varying the starting point of the simplex.

We note here that the results are quite precise for the parameter ω, while for others they are not. This is probably related to the fact that, specially close to the critical point, the statistics of $|x(t)\rangle$ is quite sensitive to variations in ω. Hence we expect $\mathcal{L}$ to have strong variations with respect to it. Likewise, for parameters whose influence on the fluctuations is less critical we expect the results of the fit to be less reliable.

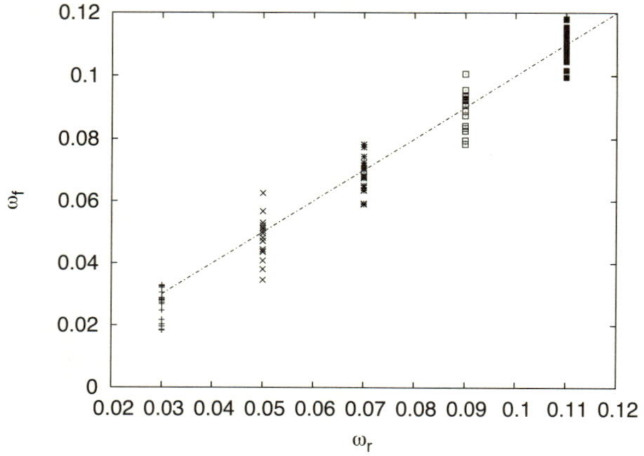

Fig. 3. Each point is a 500-days period. The data were generated with $\omega = 0.03$ (+), $\omega = 0.05$ ($\times$), $\omega = 0.07$ (Box), $\omega = 0.09$ (blacksquare). $\rho = 0.001$, $\omega_c = 0.14$, $\tau = 100$, $\epsilon = 0.1$, $\Delta = 0.04$. The starting point for the simplex was the same for all the realizations, even at different ω.

4 Fitting real markets

We now turn to the application of the algorithm to fit real data. We use daily data from 21 Dow Jones companies in the period May 1995 to May 2005 [10]. Given the discussion of the previous section, we focus on the parameters ω and ω_c. The first thing we notice is the convergence towards a stable value as we increase the length of the window T, Fig.4. The fit is done using a value of $\tau = 100$, which is a reasonable span for portfolio investment. As T increases well above τ the results get more and more stable.

Next we show the ratio ω/ω_c (inset of Fig.4), which gives a measure of the proximity to the critical point. For $\omega/\omega_c = 1$ we are at the critical point, while when this ratio is smaller we are in the stable phase.

5 Conclusions

We have presented a phenomenological model for the dynamics of inter-asset correlations in financial markets which has an interesting phase transition to an unstable phase. The model reproduces the non-trivial dynamics of the largest eigenvalue of the correlation matrix observed in real markets in the region close to the phase transition line. We have developed and discussed a maximum likelihood algorithm to fit the model parameters from empirical data of real markets. The results for Dow Jones companies confirm that real market data are best described by the model for parameters close to the critical point marking the phase transition.

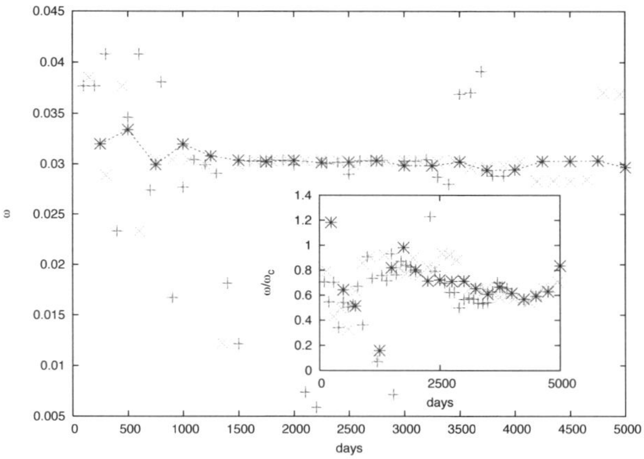

Fig. 4. Result of the fit on the Dow Jones companies. Each point is a T-days period, where $T = 200$ (+), $T = 300$ (×) and $T = 500$ ($\star$ and line). In the inset the quantity ω/ω_c is plotted for the same dataset.

A simple argument why real markets should be close to an instability is that if they were far from it, they would offer smooth and stable correlations for risk minimization strategies. This induces portfolio investment, and consequently drives the system closer to the critical point. There, the unpredictable instabilities that emerge in the market mode discourage further investment. Thus a multi-asset financial market should self-organize close to the critical point.

Our results could be generalized in a number of ways, for example extending the fitting procedure to more complex models. They can also be used in practical applications of risk management in order to estimate that component of "dynamical" risk which arises from the non-linearities induced in price dynamics by portfolio investment.

References

1. Bouchaud JP, Potters M (2003) Theory of financial risk and derivative pricing: from statistical physics to risk management, Cambridge University Press, Cambridge
2. Dacorogna MM et al (2001) An Introduction to High-Frequency Finance, Chap. V, Academic Press, London
3. Mantegna RN (1999) Eur. Phys. J. B 11:193
4. Laloux L, Cizeau P, Bouchaud JP, Potters M (1999) Phys. Rev. Lett. 83(7):1467-1470; Plerou V, Gopikrishnan P, Rosenow B, Amaral LAN, Stanley HE (1999) Phys. Rev. Lett. 83(7):1471-1474
5. Plerou V, Gopikrishnan P, Rosenow B, Amaral LAN, Stanley HE (1999) Phys. Rev. Lett. 83(7):1471-1474

6. Marsili M (2002) Quant. Fin. 2:297-302
7. Raffaelli G, Marsili M, e-print physics/0508159
8. Elton EJ, Gruber MJ (1995) Modern Portfolio theory and investment analysis. J. Wiley & sons, New York
9. $|x\rangle$ should be considered as a column vector, whereas $\langle x|$ is a row vector. Hence $\langle x|y\rangle$ is the scalar product and $|x\rangle\langle y|$ is the direct product, i.e. the matrix with entries $a_{i,j} = x_i y_j$.
10. Data was taken form `finance.yahoo.com` in the time period June 16th 1997 to May 25th 2005 for all assets except for the Dow Jones, for which we used May 2nd 1995 to May 23rd 2005. Correlations were measured on the set of assets composing the index at the final date.
11. Raffaelli G, Ponsot B, Marsili M., in preparation.

Agents Play Mix-game

Chengling Gou

Physics Department, Beijing University of Aeronautics and Astronautics, 37 Xueyuan Road, Heidian District, Beijing, China, 100083
gouchengling@buaa.edu.cn

1 Introduction

In recent years, economics and finance see the shift of paradigm from representative agent models to heterogeneous agent models [1, 2]. More and more economists and physicists made efforts in research on heterogeneous agent models for financial markets. Minority game (MG) proposed by D. Challet, and Y. C. Zhang [3] is an example among such efforts. Challet and Zhang's MG model, together with the original bar model of Arthur, attracts a lot of following studies [4–6] . Given MG's richness and yet underlying simplicity, MG has also received much attention as a financial market model [4]. MG comprises an odd number of agents choosing repeatedly between the options of buying (1) and selling (0) a quantity of a risky asset. The agents continually try to make the minority decision, i.e. buy assets when the majority of other agents are selling, and sell when the majority of other agents are buying. Neil F. Johnson [4, 5] and coworkers extended MG by allowing a variable number of active traders at each timestep— they called their modified game as the Grand Canonical Minority Game (GCMG). GCMG, and to a lesser extent the basic MG itself, can reproduce the stylized facts of financial markets, such as volatility clustering and fat-tail distributions.

However, there are some weaknesses in MG and GCMG. First, the diversity of agents is limited, since agents all have the same history memory and time-horizon. Second, in real markets, some agents are trend-followers, i.e. 'noise traders' [7–13, 19], who effectively play a majority game; while others are 'fundamentalist', who effectively play a minority game. De Martino, I. Giardinal, M. Marsili and A. Tedeschi have done some research on mixed Minority/Majority Games analytically and simulationally [6, 14–17] . In their models, time-horizon is infinite and all agents have the same ability to deal with information and can switch between majority and minority. They study the stationary macroscopic properties and the dynamical features of the systems under different information structures, i.e. agents receive the information which is either exogenous ('random') or endogenous ('real'). They find that

(a) a significant loss of informational efficiency with respect to a pure minority game, in particular, an efficient and unpredictable phase exists only if the fraction of agents in majority group $< 1/2$, and (b) a catastrophic increase of global fluctuations if the fraction of agents in majority group $> 1/2$ under the condition that agents receive random information [15]. They also find that a small amount of herding tendency can alter the collective behavior dramatically if agents receive endogenous information (real history) [16]. Zhong et al study a similar model in which they introduce contrarians into MG who deliberately prefer to hold an opinion that is contrary to the prevailing idea of the commons or normal agents. Their results of numerical simulations reveal that the average success rate among agents depends non-monotonically on the fraction of contrarians [18].

In order to create an agent-based model which more closely mimics a real financial market, this paper modifies MG model by dividing agents into two groups: each group has different history memory and time-horizon; one group plays the minority game and the other plays the majority game. For this reason, this system refers to as a 'mix-game' model. The difference between mix-game and the mixed Minority/Majority Games studied by Marsili and Martino et al is that the two groups of agents in mix-game have different bounded abilities to deal with history information and to count their own performances. This feature of mix-game represents that agents have different bounded rationality [19] . Reference [20] reported that almost all agents play with history memories of 6 or less in a typical minority game with evolution. This means history memories of 6 or less is more important and worthwhile to look at. Therefore, I focus on studying the effect on MG by adding some agents playing majority game with history memories of 6 or less.

This paper focuses on studying the difference of the dynamic behaviors of mix-game model under the conditions of different configuration of agent history memories with agents receiving real history information. It looks at the effect on local volatilities by increasing the number of agents who play majority game and the effect on local volatilities by changing time horizons of the two groups. And it also studies the relations between local volatilities and average individual performance of agents caused by changing the history memories of agents. Section 2 introduces mix-game model and the simulation condition. Section 3 reports the simulation results and discussions. Section 4 gives an brief idea of application of mix-game model. Section 5 reaches the conclusion of this paper.

2 Mix-game model and simulation condition

In mix-game, there are two groups of agents; group 1 plays the majority game, and the group 2 plays the minority game. N is the total number of the agents and N_1 is number of agents in group 1. The system resource is $R = 0.5 \times N$. All agents compete in the system for the limited resource R. T_1 and T_2 are time horizons of the two groups of agents, and m_1 and m_2 denote the history memories of the two groups of agents, respectively.

Only the global information available to the agents is a common bit-string 'memory' of m_1 or m_2 most recent competition outcomes. A strategy consists of a response, i.e., 0 or 1, to each possible bit string; hence there are $2^{2^{m_1}}$ or $2^{2^{m_2}}$ possible strategies for group 1 or group 2, respectively, which form full strategy spaces (FSS). At the beginning of the game, each agent is assigned s strategies and keeps them unchangeable during the game. After each turn, agents assign one virtual point to a strategy which would have predicted the correct outcome. For agents in group 1, they will reward their strategies one point if they are in the majority; for agents in group 2, they will reward their strategies one point if they are in the minority. Agents collect the virtual points for their strategies over the time horizon T_1 or T_2, and agents use their strategy which has the highest virtual point in each turn. If there are two strategies which have the highest virtual point, agents use coin toss to decide which strategy to be used. Excess demand $D[t^-]$ is equal to the number of ones (buy) which agents choose minus the number of zeros (sell) which agents choose.

$$D[t^-] = n_{(buy-orders)}[t-1] - n_{(sell-orders)}[t-1] \tag{1}$$

According to a widely accepted assumption that excess demand exerts a force on the price of the asset and the change of price is proportional to the excess demand in a financial market [21,22], the time series of price of the asset $P(t)$ can be calculated based on the time series of excess demand.

$$P[t] = D[t^-]/\lambda + P[t-1] \tag{2}$$

Volatility of the time series of prices is represented by the variance of the increases of prices. Local volatility (Vol) is calculated at every time-step by calculating the variance of the increases of prices within a small time-step window d. If $\lambda = 1$ chosen for simplicity, the increase of prices is just the excess demand. Therefore, we get formula (3).

$$Vol[t] = \frac{1}{d}\sum_{t-d}^{t} D[t^-]^2 - (\frac{1}{d}\sum_{t-d}^{t} D[t^-])^2 \tag{3}$$

In simulations, the distribution of initial strategies of agents is randomly uniform in FSS and remains unchanged during the game. Each agent has two strategies, i.e. $s = 2$. The simulation running time-steps are 3000. The window length of local volatility is $d = 5$.

3 Simulation results and discussions

In order to see the difference of the dynamic behaviors of mix-game model under the conditions of different configuration of agent history memories of the two groups, I choose three categories of configurations of history memories: (a) $m_1 < m_2 = 6$; (b) $m_1 = m_2 \leq 6$; (c) $m_1 = 6 > m_2$.

3.1 Effects on local volatilities caused by increasing agents in majority group

To see the effect of mix-game on local volatilities caused by increasing agents in majority group, simulations are done with different parameter configurations of history memories, different number of total agents and different fraction of agents in group 1. Number of total agents (N) varies from 101 to 401. Time horizons (T_1 and T_2) are relatively stable. Three typical results are shown in Fig.1.

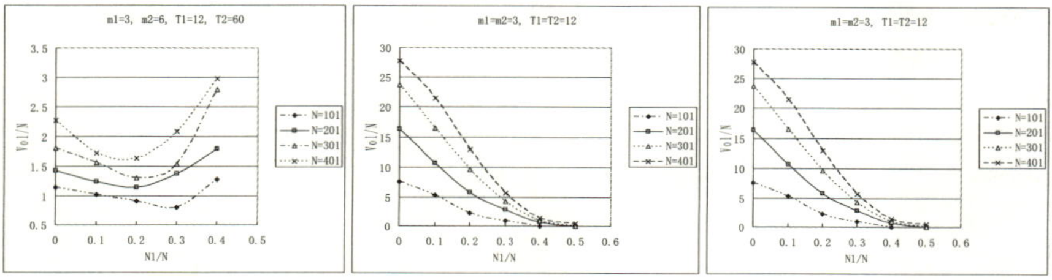

Fig.1a Fig.1b Fig.1c

Fig. 1. Vol/N changes with N_1/N under the condition of different parameter configurations of history memories and N from 101 to 401, where Vol represents mean of local volatilities

For the history memory configuration of $m_1 < m_2 = 6$, simulations have been done under the conditions of (a) $m_1 = 1, m_2 = 6, T_1 = 12, T_2 = 60$; (b) $m_1 = 3, m_2 = 6, T_1 = 12, T_2 = 60$; (c) $m_1 = 5, m_2 = 6, T_1 = 12, T_2 = 60$. The results are similar in these three conditions. Fig.1a shows one typical example of simulation results about the means of local volatilities per agent (Vol/N) change when N_1/N increases from 0 to 0.4 under the condition of $m_1 = 3, m_2 = 6, T_1 = 12, T_2 = 60$ and $N = 101, 201, 301, 401$. We can find that the means of local volatilities per agent have the lowest points at about $N_1/N = 0.2$ or 0.3. Reference [18] found similar results when contrarians are introduced into MG, but the difference is that the lowest point of volatility appears when the fraction of contrarians is 0.4.

For the history memory configuration of $m_1 = m_2 \leq 6$, simulations have been done under the conditions of (a) $m_1 = 1, m_2 = 1, T_1 = 12, T_2 = 12$; (b) $m_1 = 3, m_2 = 3, T_1 = 12, T_2 = 12$; (c) $m_1 = 6, m_2 = 6, T_1 = 12, T_2 = 12$. The results are similar in these three conditions. Fig.1b shows one typical example of simulation results about the means of local volatilities per agent(Vol/N)

change when N_1/N increases from 0 to 0.4 under the condition of $m_1 = m_2 = 3, T_1 = T_2 = 12$ and $N = 101, 201, 301, 401$. From Fig.1b we can find that the means of local volatilities per agent decrease and approach to 1 while N_1/N increase from 0 to 0.4. This means the efficiency of a system can be improved by adding some agents with the same parameters (i.e. $m_1 = m_2, T_1 = T_2$.) who play the majority game. Reference [15] reported the similar results under the condition of random information with $N_1/N < 0.5$. And reference [23] observed the similar phenomena in distributed computing systems.

Fig.1c shows the simulation results about the means of local volatilities per agent (Vol/N) change when N_1/N increases from 0 to 0.4 with $m_1 = 6, m_2 = 3$ and $T_1 = 60, T_2 = 12$. From Fig. 1c, we can notice that the means of local volatilities per agent have decreasing tendencies while N_1/N increases from 0 to 0.4. It is also found that the means of local volatilities per agent decrease when N_1/N increases from 0 to 0.4 for situations of $m_1 = 6, m_2 = 1, T_1 = 60, T_2 = 12$, which is similar to that shown in Fig.1c, but the mean of local volatilities per agent (Vol/N) have the lowest points at $N_1/N = 0.1$ for situation of $m_1 = 6, m_2 = 5, T_1 = 60, T_2 = 12$.

From Fig.1, we also can notice that the means of local volatilities per agent increase while N increases from 101 to 401 in all situations. Through simulations, I find that local volatilities drop to zeros in most cases when N_1/N is larger than 0.5 under the simulation conditions of $m_1 = m_2$ or $m_1 < m_2 = 6$ with some initial strategy distributions. This phenomenon also is influenced by T_1. But this phenomenon has not been found under the simulation condition of $m_1 = 6 > m_2$. The reason about local volatilities dropping to zeros is that all agents quickly adapt to using one strategy and stick to it, i.e. all agents are 'frozen' in these simulation conditions.

3.2 Dynamics of mix-game with different time horizons

Finite time horizons for MG have influence on the dynamics of price time series, as studied by Hart and Johnson [25] and D. Chalet [26].To look at the influence of time horizons (T_1, T_2) on dynamic behaviors of mix-game models, simulations have been done under three categories as mentioned in the beginning of this section. It is found that the dynamic behaviors of mix-game are similar under the conditions of $m_1 < m_2 = 6$ when T_1 or T_2 increases with different m_1. One example of such simulations is shown in Fig.2 which plots the relations between means of local volatilities and time horizon T_1 or T_2) when $T_2 = 36$ or $T_1 = 36, m_1 = 3, m_2 = 6, N = 201, N_1 = 72$ and $s = 2$. From Fig.2, one can see that means of local volatilities are stable if $T_1 < T_2$; otherwise, the means of local volatilities are unstable and larger than the stable values if $T_1 > T_2$. In contrast, under the condition of $m_1 = 6 > m_2$, the dynamic behaviors of mix-game are quit different from that under the condition of $m_1 < m_2 = 6$ when T_2 increase with fixed T_1. One example is shown in Fig.3 which plots the relations between means of local volatilities and time horizon T_2 when $T_1 = 36, m_1 = 6, m_2 = 3, N = 201, N_1 = 72$

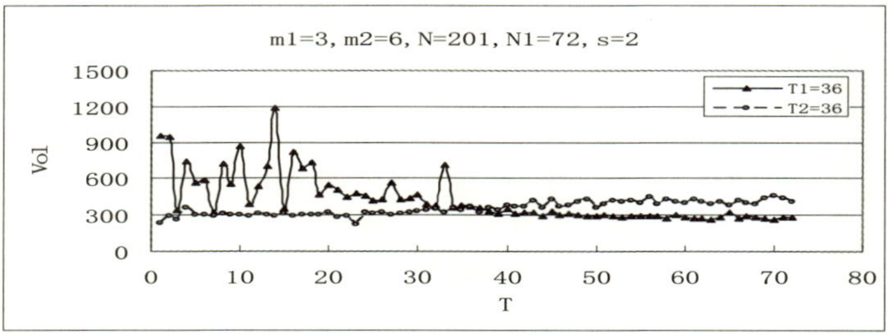

Fig. 2. Relations between means of local volatilities and time horizon T_1 or T_2 when $T_2 = 36$ or $T_1 = 36, m_1 = 3, m_2 = 6, N = 201, N_1 = 72$ and $s = 2$

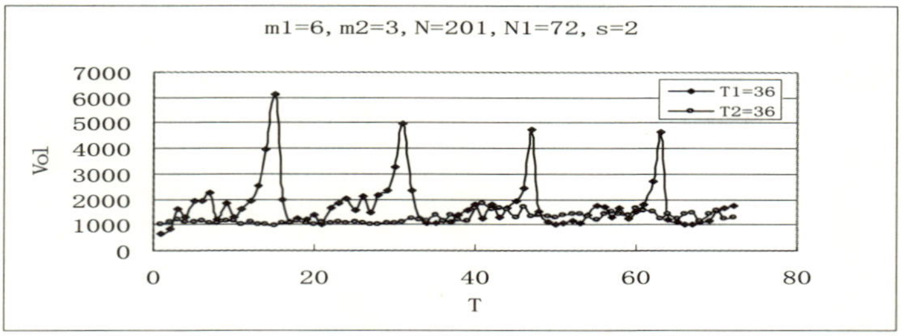

Fig. 3. Relations between means of local volatilities and time horizon T_1 or T_2 when T_2 or $T_1 = 36, m_1 = 6, m_2 = 3, N = 201, N_1 = 72$ and $s = 2$

and $s = 2$. From Fig.3, one can notice that periodic phenomenon of local volatilities with respect to T_2 whose period is 2×2^{m_2}.This result is similar to that discovered by Hart and Johnson when they studied MG with finite time horizon [25]. This implies that agents in group 1 have little influence on the behaviors of agents in group 2 under the condition of $m_1 = 6 > m_2$. The relation between means of local volatilities and time horizon T_1 when $T_2 = 36, m_1 = 6, m_2 = 3, N = 201, N_1 = 72$ and $s = 2$ is also shown in Fig.3. When $T_1 < T_2$, the means of local volatility are relatively stable. But when $T_1 > T_2$, the means of local volatilities with respect to T_1 are more fluctuating.

Fig.4 shows the relations between means of local volatilities and time horizon

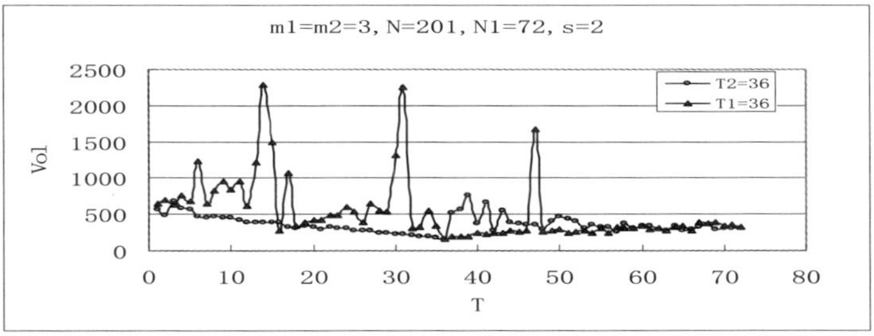

Fig. 4. Relations between means of local volatilities and time horizon T_1 or T_2 when T_2 or $T_1 = 36, m_1 = m_2 = 3, N = 201, N_1 = 72$ and $s = 2$

T_1 or T_2 when T_2 or $T_1 = 36, m_1 = m_2 = 3, N = 201, N_1 = 72$ and $s = 2$ under the condition of $m_1 = m_2 = 3$. Comparing Fig.4 with Fig.3, we can notice three differences about the relations between means of local volatilities and time horizon T_2. First, there are only three peaks in Fig.4; second, the heights of the peaks in Fig.4 are much smaller than that in Fig.3; third, the first peak in Fig.4 is at $T_2 = 14$ while the first peak in Fig.3 is at $T_2 = 15$. From Fig.4, we also can find that the means of local volatilities are relatively stable if $T_1 < T_2$ which is similar to that shown in Fig.2.

3.3 Correlations between average winnings of each group and means of local volatilities

To look at the correlations between volatilities of price time series and average winning of each group, simulations have been done with parameters of $T_1 = T_2 = 12, N = 201, N_1 = 72$ and the three configurations of history memories as specified in the beginning of this section, respectively.

Through simulation, it is found that both the means of local volatilities and the average winnings of these two groups increase when m_1 decrease from 6 to 1 if $m_1 < m_2 = 6$. This implies that agents with smaller memory lengths in group 1 can not only improve their own performance but also benefit for agents in group 2. The cooperation between group 1 and group 2 emerges in this situation. Table 1 shows that correlations among R_1, R_2 and Vol_1 are largely positive, where R_1 and R_2 represent the average winning of group 1 and group 2, respectively, and Vol_1 represents the means of local

Table 1. Correlations among R_1, R_2 and Vol_1 when $m_2 = 6$, and m_1 increases from 1 to 6

	R_1	R_2	Vol_1
R_1	1		
R_2	0.982	1	
Vol_1	0.985	0.996	1

volatilities. These results mean that the improvement of the performance of individual agents accompanies with the decrease of the efficiency of systems under this simulation condition. Agents can benefit from the larger fluctuation of systems. This is accordant with the reality of financial markets. Therefore, mix-game models with this category of history memory configurations can be considered as candidates for modelling financial markets.

Table 2. Correlations among R_1, R_2 and Vol_2 when $m_1 = 6$, and m_2 increases from 1 to 6

	R_1	R_2	Vol_2
R_1	1		
R_2	-0.246	1	
Vol_2	0.148	-0.892	1

If $m_1 = 6 > m_2$, the average winning of group 2 decreases, but the average winning of group 1 does not change greatly, while the mean of local volatilities increase when m_2 decreases from 6 to 1. Table 2 shows that the average winning of group 2 (R_2) strongly negatively correlated with the volatilities of systems (Vol_2), where R_1 and R_2 represent the average winning of group 1 and group 2, respectively, and Vol_2 represents the means of local volatilities. But the average winning of group 1 (R_1) just slightly positively correlated with the volatilities of systems (Vol_2). This result is due to the uncorrelated strategy spaces between these two groups [27].

Table 3. Correlations among R_1, R_2 and Vol_3 when $m_1 = m_2$, and they increase from 1 to 6

	R_1	R_2	Vol_3
R_1	1		
R_2	0.977	1	
Vol_3	-0.859	-0.737	1

If $m_1 = m_2$, the average winnings of these two groups decrease, but the mean of local volatilities increases when both m_1 and m_2 decreases from 6 to 1. Table 3 shows that the average winnings of group 1 and group 2 (R_1 and R_2) greatly negatively correlated with the volatilities of systems (Vol_3), where R_1 and R_2 represent the average winning of group 1 and group 2, respectively, and Vol_3 represents the means of local volatilities. In this situation, average winnings of these two groups are similar to that of MG, but the volatility decreases greatly just due to their anticorrelated strategy spaces [27].

4 Application of mix-game model

From the former section, one can find out that mix-game with some parameters reproduces the stylized features of financial time series, but some fail to do so. Therefore, one needs to choose parameters of m_1, m_2, T_1, T_2, N and N_1 by using mix-game to model financial markets. The following aspects need to be considered:

- First make sure the time series of mix-game can reproduce the stylized facts of price time series of financial markets by choosing proper parameters: $m_1 < m_2 = 6, T_1 < T_2, N1/N < 0.5$;
- Second pay attention to the fluctuation of local volatilities, and ensure that the median of local volatilities of mix-game is similar to that of the target time series;
- Third make sure the log-log plot of absolute returns look similar.

Since the median value of local volatilities of Shanghai Index daily data from 1992/01/02 to 2004/03/19 is 222, one combination of parameters of mix-game has the similar median values of local volatilities according to Fig.1, which is $m_1 = 3, m_2 = 6, T_1 = 12, T_2 = 60, N = 201, N_1 = 40$ [28].

5 Summary and conclusions

In mix-game, the local volatilities, the average winnings of agents and the correlations between the average winnings of agents and the means of local volatilities largely depend on the configurations of agent history memories and the fraction of agents in group 1. Volatilities are also influenced by time horizons of these two groups and such influences also depend on the configurations of the configurations of agent history memories.

The underlying mechanisms for the above findings are as following: if these two groups have the same history memories, their strategy spaces are anticorrelated; if history memory of group 1 is 6 and larger than that of group 2, their strategy spaces are uncorrelated; if history memory of group 2 is 6 and larger than that of groups 1, their strategy spaces are partly anticorrelated. The detail analysis can be found in reference [27].

Mix-game can be a potentially good model for financial markets with specific parameters. For Shanghai Index, there is one suitable configuration of parameters: $m_1 = 3, m_2 = 6, T_1 = 12, T_2 = 60, N = 201, N_1 = 40$. For any other financial markets, parameters need to be adjusted.

References

1. Levy M, Levy H, Solomon S (1994) Economics Letters 45: 103–111
2. Levy M, Levy H, Solomon S (2000) Microscopic Simulation of Financial Markets. Academic Press, New York
3. Challet D, Zhang Y C (1997) Phyisca A 246: 407
4. Johnson N F, Jefferies P, Hui P M (2003), Financial Market Complexity. Oxford University Press, Oxford
5. Jefferies P, Johnson N F (2001) Oxford Center for Computational Finance working paper: OCCF/010702
6. Coolen A C C (2005) The Mathematical Theory of Minority Games. Oxford University Press, Oxford
7. Andersen J V, Sornette D (2003) Eur. Phys. J. B 31: 141–145
8. Lux T (1995) Economic Journal 105(431): 881–896
9. Lux T, Marchesi M (1999) Nature 397(6719): 498–500
10. Challet D (2005) arXiv: physics/0502140 v1
11. Slanina F, Zhang Y-C (2001) Physica A 289: 290
12. Yoon S M, Kim K (2005) arXiv: physics/0503016 v1
13. Giardina I, Bouchaud J P (2003) Eur. Phys. J. B 31: 421
14. Marsili M (2001) Physica A 299: 93
15. Martino A D, Giardina I, Mosetti G (2003) J. Phys. A 36: 8935
16. Tedeschi A, Martino A D, Giardina I(2005) arXiv: cond-mat/0503762
17. Martino A D, Giardinal I, Marsili M, Tedeschi A (2004) arXiv: cond-mat/0403649
18. Zhong L X, Zheng D F,Zheng B, Hui P M (2004) arXiv: cond-mat/0412524
19. Shleifer A (2000) Inefficient Markets: an Introduction to Behavioral Financial. Oxford University Press, Oxford
20. Savit R, Koelle K, Treynor W, Gonzalez R (2004) In: Tumer K, Wolpert D (eds) Collectives and the Desing of Complex System. Springer-Verlag, New York , P199 212
21. Bouchaud J P, Cont R (1998) Eur. Phys. J. B 6: 543
22. Farmer J D (2002) Industrial and Corporate Change 11: 895-953
23. Kephart J O, Hogg T, Huberman B A (1989) Physical Review A 40(1): 404-421
24. Yang C (2004) Thesis of Beijing University of Aeronautics and Astronautics, Beijing
25. Hart M L, Johnson N F(2002) Physica A 311: 275 – 290
26. Chalet D (2004) arXiv: cond-mat/0407595.
27. Gou C (2005) arXiv: physics/0504001 v3, accepted by Chinese Physics
28. Gou C (2005) www.cfrn.com.cn: paperID=1548, submitted to JASSS

Triangular Arbitrage as an Interaction in Foreign Exchange Markets

Yukihiro Aiba[1] and Naomichi Hatano[2]

[1] Department of Physics, University of Tokyo, Komaba, Meguro, Tokyo 153-8505, Japan. aiba@iis.u-tokyo.ac.jp
[2] Institute of Industrial Science, University of Tokyo, Komaba, Meguro, Tokyo 153-8505, Japan. hatano@iis.u-tokyo.ac.jp

1 Introduction

Analyzing correlation in financial time series is a topic of considerable interest [1–17]. In the foreign exchange market, a correlation among the exchange rates can be generated by a triangular arbitrage transaction. The purpose of this article is to review our recent study [18–23] on modeling the interaction generated by the triangular arbitrage.

The triangular arbitrage is a financial activity that takes advantage of the three exchange rates among three currencies [24, 25]. Suppose that we exchange one US dollar to some amount of Japanese yen, exchange the amount of Japanese yen to some amount of euro, and finally exchange the amount of euro back to US dollar; then how much US dollar do we have? There are opportunities that we have more than one US dollar. The triangular arbitrage transaction is the trade that takes this type of opportunities.

In order to quantify the triangular arbitrage opportunities, we define the quantity

$$\nu(t) = \ln \prod_{x=1}^{3} r_x(t) = \sum_{x=1}^{3} \ln r_x(t). \tag{1}$$

where $r_x(t)$ denotes each exchange rate at time t. We refer to this quantity as the logarithm rate product. There is a triangular arbitrage opportunity whenever the logarithm rate product is positive: $\nu > 0$. Once there is a triangular arbitrage opportunity, many traders will make the transaction. This makes ν converge to a slightly negative value [25], thereby eliminating the opportunity; the triangular arbitrage is thus a form of interaction among currencies. Triangular arbitrage opportunities nevertheless appear, because each rate r_x fluctuates strongly.

For those who might doubt that triangular arbitrage transactions are carried out in actual markets, we can think of the following transaction. Suppose that a Japanese company earned some amount of US dollar and wants to exchange it to Japanese yen. There can be instances when it is more advantageous to exchange the US dollar to Japanese yen *via euro* than to exchange it directly to Japanese yen. The condition for such instances to happen is again $\nu > 0$ and the actual occurrence of such transactions again makes the logarithm rate product converge to zero.

Figure 1 is a semi-logarithmic plot of the probability density function of the logarithm rate product ν. It has a sharp peak around $\epsilon \equiv \langle \nu \rangle \simeq -0.00091$ and fat tails; it is far from a Gaussian distribution. (The average is slightly negative because of the spread; the spread is the difference between the ask and the bid prices and is usually of the order of 0.05% of the prices.) This suggests that the triangular arbitrage indeed affects the fluctuation in foreign exchange markets.

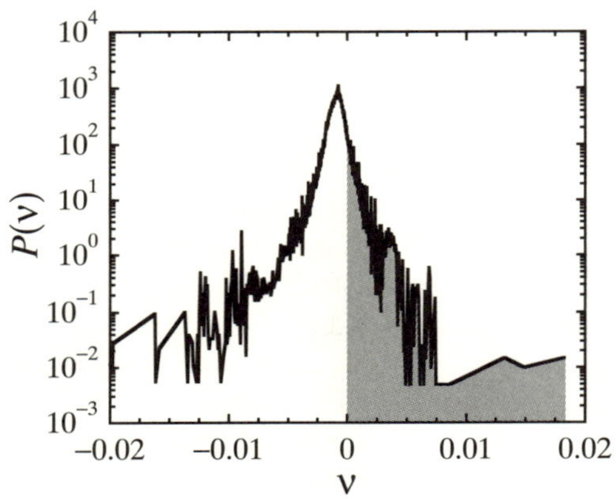

Fig. 1. The probability density function of the logarithm rate product ν. The shaded area represents triangular arbitrage opportunities. The data were taken from January 25 1999 to March 12 1999.

We here present our macroscopic and microscopic models that describe an effect of triangular arbitrage transactions on multiple markets. Many models of price change have been introduced so far: for example, the Lévy-stable non-Gaussian model [1]; the truncated Lévy flight [26]; the ARCH/GARCH processes [27,28]. They discuss, however, only the change of one price and do not consider an interaction among multiple prices.

2 A macroscopic model of triangular arbitrage transaction

The basic equation of our macroscopic model is a time-evolution equation of the logarithm of each rate [18–21]:

$$\ln r_x(t+T) = \ln r_x(t) + \eta_x(t) + g(\nu(t)) \qquad (x = 1, 2, 3), \tag{2}$$

where ν is the logarithm rate product (1), and T is a time step which controls the time scale of the model; we later use the actual financial data every T[sec]. Equation (2) attributes the change of the logarithm of each rate to an independent fluctuation $\eta_x(t)$ and an attractive interaction

$$g(\nu) \begin{cases} < 0 & \text{for } \nu > \epsilon, \\ > 0 & \text{for } \nu < \epsilon, \end{cases} \tag{3}$$

where $\epsilon \equiv \langle \nu \rangle \simeq -0.00091$. As a linear approximation, we define $g(\nu)$ as

$$g(\nu) \equiv -k(\nu - \epsilon) \tag{4}$$

where k is a positive constant which specifies the interaction strength.

The time-evolution equation of ν is given by summing Eq. (2) over all x:

$$\nu(t+T) - \epsilon = (1 - 3k)(\nu(t) - \epsilon) + F(t), \tag{5}$$

where $F(t) \equiv \sum_{x=1}^{3} \eta_x(t)$. This is our basic time-evolution equation of the logarithm rate product.

From a physical point of view, we can regard the model equation (2) as a one-dimensional random walk of three particles with a restoring force, by interpreting $\ln r_x$ as the position of each particle. The logarithm rate product ν is the summation of $\ln r_x$, hence is proportional to the center of gravity of the three particles. The restoring force $g(\nu)$ makes the center of gravity converge to a certain point $\epsilon = \langle \nu \rangle$. The form of the restoring force (4) is the same as that of the harmonic oscillator. Hence we can regard the coefficient k as a spring constant.

The spring constant k is related to the auto-correlation function of ν as follows:

$$1 - 3k = c(T) \equiv \frac{\langle \nu(t+T)\nu(t) \rangle - \langle \nu(t) \rangle^2}{\langle \nu^2(t) \rangle - \langle \nu(t) \rangle^2}. \tag{6}$$

Using Eq. (6), we can estimate k from the real data series as a function of the time step T. The spring constant k increases with the time step T. We hereafter fix the time step at $T = 60$[sec] and use $k(1[\text{min}]) = 0.17 \pm 0.02$ for our simulation. We will come back to this point in Sect. 4.

On the other hand, the fluctuation of a foreign exchange rate is known to be a fat-tail noise [29,30]. Here we take $\eta_x(t)$ as the truncated Lévy process [**?**,26] and determine the parameters from the real market data. We thereby simulated the time evolution (5). The probability density function of the results (Fig. 2) is compared to that of the real data in Fig. 1. The fluctuation of the simulation data is consistent with that of the real data. In particular, we see good agreement around $\nu \simeq \epsilon$ as a result of the linear approximation of the interaction function.

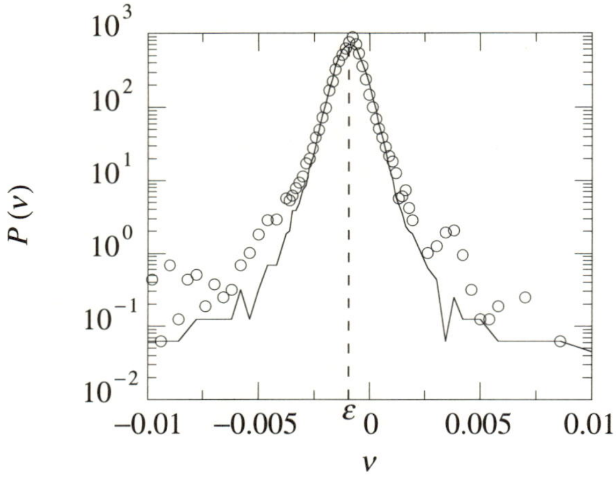

Fig. 2. The probability density function of ν. The circle (∘) denotes the real data and the solid line denotes our simulation data with the interaction. The simulation data fit the real data well.

3 A microscopic model of triangular arbitrage transaction

The model (2) that we introduced in the previous section is phenomenological; i.e. it treats the fluctuations of the rates as fluctuating particles and the interaction among the rates as a spring. We refer to this model as the 'macroscopic model' hereafter. We now introduce another of our models which focuses on the dynamics of each dealer in the markets [23]; we refer to the new model as the 'microscopic model' hereafter.

In order to describe each foreign exchange market microscopically, we use Sato and Takayasu's dealer model (the ST model) [32], which reproduces the power-law behavior of price changes in a single market well. Although we focus on the interactions among three currencies, two of the three markets can be regarded as one effective market [21]; i.e. the yen-euro rate and the euro-dollar rate are combined to an effective yen-dollar rate. This means that each dealer is put in the situation of the Japanese company that we mentioned in Sect. 1. He/she wants to buy and sell yen and dollar sometimes directly and sometimes via euro. We therefore describe triangular arbitrage opportunities with only two interacting ST models, in order to simplify the situation. We refer to the two ST models as the market X and the market Y.

The basic assumption of the ST model is that dealers want to buy stocks or currencies at a lower price and to sell them at a higher price. There are N dealers in a market x; the ith dealer has bidding prices to buy, $B_{i,x}(t)$, and to sell, $S_{i,x}(t)$, at time t. We assume that the difference between the buying price and the selling price is a constant $\Lambda \equiv S_{i,x}(t) - B_{i,x}(t) > 0$ for all i and x, in order to simplify the model.

An intra-market transaction takes place when the condition [32]

$$\max\{B_{i,x}(t)\} \geq \min\{S_{i,x}\} \quad (x = X \text{ or } Y) \tag{7}$$

is satisfied. The price $P_x(t)$ of the market x is renewed in analog to the ST model [32]:

$$P_x(t) = \begin{cases} (\max\{B_{i,x}(t)\} + \min\{S_{i,x}(t)\})/2 & \text{if the condition (7) is satisfied,} \\ P_x(t-1) & \text{otherwise,} \end{cases} \tag{8}$$

where $x = X$ and Y.

We here add a new *inter*-market transaction rule which makes the systems interact. When a dealer cannot find another dealer to transact within his/her own market (or to exchange dollar to yen directly), he/she tries to find one in the other market (or to exchange dollar to yen via euro). An arbitrage transaction can take place when the condition

$$\nu_x \equiv \max\{B_{i,x}(t)\} - (\min\{B_{i,\overline{x}}(t)\} + \Lambda) \geq 0 \tag{9}$$

is satisfied, where $(x, \overline{x}) = (X, Y)$ or (Y, X). When the conditions (7) and (9) (for either of $x = X$ or Y) are both satisfied simultaneously, the condition (7) precedes.

Note that the arbitrage conditions $\nu_X \geq 0$ and $\nu_Y \geq 0$ in the microscopic model correspond to the arbitrage condition $\nu \geq 0$ in the actual market, where ν is defined by Eq. (1). We assume that the dealers' bidding prices $\{B_{i,x}\}$ and $\{S_{i,x}\}$ correspond to the logarithm of the exchange rate, $\ln r_x$. Therefore, $\max\{B_{i,X}\}$ may be equivalent to $-\ln$(yen-dollar ask) while $\min\{S_{i,Y}\}$ may be equivalent to $\ln$(dollar-euro ask) $- \ln$(yen-euro bid), and hence ν_X may be equivalent to ν.

The dealers in the markets X and Y change their bidding prices according to the deterministic rule of the ST model [32]:

$$B_{i,x}(t+1) = B_{i,x}(t) + a_{i,x}(t) + c\Delta P_x(t) \tag{10}$$

where $a_{i,x}(t)$ denotes each dealer's characteristic movement in the price at time t, $\Delta P_x(t)$ is the difference between the price at time t and the price at the time when the previous trade was done, and $c(> 0)$ is a constant which specifies dealers' response to the market price change and is common to all of the dealers in each market. The absolute value of a dealer's characteristic movement $a_{i,x}(t)$ is given by a uniform random number in the range $[0, \alpha)$ and is fixed throughout the time. The sign of $a_{i,x}$ is positive when the dealer is a buyer and is negative when the dealer is a seller. The buyer (seller) dealers move their prices up (down) until any of the conditions (7) or (9) is satisfied. Once the transaction takes place, the buyer of the transaction becomes a seller and the seller of the transaction becomes a buyer; in other words, the buyer dealer changes the sign of $a_{i,x}$ from positive to negative and the seller dealer changes it from negative to positive.

The initial values of $\{B_{i,x}\}$ are given by uniform random numbers in the range $(-\Lambda, \Lambda)$. We thus simulate this model specifying the following four parameters: the number of dealers, N; the spread between the buying price and the selling price, Λ; the dealers' response to the market price change, c; and the average of dealers' characteristic movements in a unit time, α. The procedures of the simulation of the microscopic model are as follows:

1. Prepare two systems of the ST model, the market X and the market Y. The parameters are common to the two systems.
2. Check the condition (7) for each market. If the condition (7) is satisfied, renew the price $P_x(t)$ by Eq. (8), skip the step 3 and proceed to the step 4. Otherwise, proceed to the step 3.
3. Check the arbitrage condition (9). If the condition (9) is satisfied for either $x = X$ or Y, renew the prices $P_x(t)$ and $P_{\overline{x}}(t)$ to $\max\{B_{i,x}(t)\}$ and $\min\{B_{i,\overline{x}}(t)\} + \Lambda$, respectively. If the condition (9) is satisfied for both $x = X$ and Y, choose one of them with the probability of 50% and carry out the arbitrage transaction as described just above. If the arbitrage transaction takes place, proceed to the step 4; otherwise skip the step 4 and proceed to the step 5.
4. Calculate the difference between the new prices and the previous prices, $\Delta P_x(t) = P_x(t) - P_x(t-1)$.
5. If any of the conditions (7) and (9) are not satisfied, maintain the previous prices, $P_x(t) = P_x(t-1)$, as well as the previous price differences, $\Delta P_x(t) = \Delta P_x(t-1)$.
6. Change the dealers' bidding prices following Eq. (10).
7. Change the buyer and the seller of the transaction to a seller and a buyer, respectively. In other words, change the signs of $a_{i,x}$ of the dealers who transacted.
8. Repeat the steps from 2 to 7.

The quantities ν_X and ν_Y are shown in Fig. 3. The fat-tail behavior of the price difference ν_X is consistent with the actual data as well as with the macroscopic model. Furthermore, ν_X and ν_Y reproduces the skewness of the actual data, which cannot be reproduced by the macroscopic model.

4 The microscopic parameters and the macroscopic spring constant

In this section, we discuss the relation between the macroscopic model and the microscopic model through the interaction strength, or the spring constant k. In the microscopic model, we define the spring constant k_{micro}, which corresponds to the spring constant k of the macroscopic model, as follows:

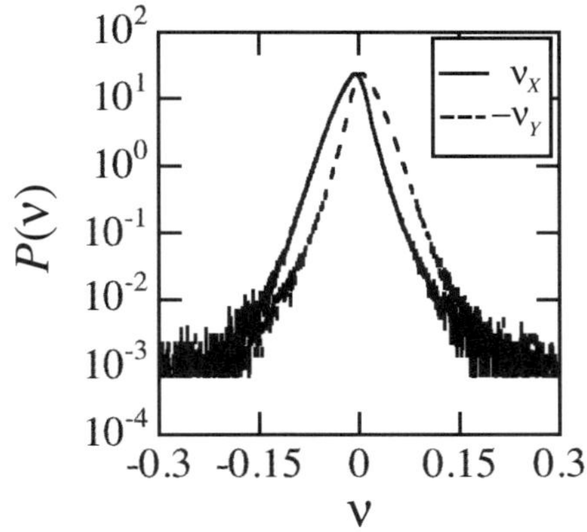

Fig. 3. The distributions of ν_X and $-\nu_Y$. The parameters are fixed to $N = 100$, $\alpha = 0.01$, $\Lambda = 1.0$ and $c = 0.3$, which follows Ref. [32], and are common to the market X and the market Y. The solid line denotes ν_X and the dashed line denotes ν_Y in each graph.

$$k_{\text{micro}} \equiv \frac{1}{2}\left(1 - \frac{\langle \nu_X(t+1)\nu_X(t)\rangle - \langle \nu_X(t)\rangle^2}{\langle \nu_X(t)^2\rangle - \langle \nu_X(t)\rangle^2}\right). \tag{11}$$

Remember that, in the macroscopic model, the spring constant k depends on the time step T. The spring constant of the microscopic model k_{micro} also depends on a time scale as follows. The time scale of the ST model may be given [32] by the following combination of the parameters $\langle n\rangle \simeq 3\Lambda/N\alpha$, where n denotes the interval between two consecutive trades. Hence, the inverse $f \equiv 1/\langle n\rangle \simeq N\alpha/3\Lambda$ is the frequency of the trades.

Although there are four parameters N, α, Λ and c, we change only three parameters N, α, and c and set $\Lambda = 1$, because only the ratios N/Λ and α/Λ are relevant in this system. The ratio N/Λ controls the density of the dealers and α/Λ controls the speed of the dealers' motion on average.

We plot the spring constant k_{micro} as a function of the trade frequency $f \equiv N\alpha/3\Lambda$ in Fig. 4(a). The plots show that the spring constant $k_{\text{micro}}(N, \alpha, \Lambda)$ can be scaled by the trade frequency f well. We argued that the data in the region $1/3 \le f \le 2/3$ are reasonable to use; see Ref. [23] for details.

The result should be compared to Fig. 4(b), where we plotted the spring constant k defined in Eq. (6) as a function of the trade frequency f_{real}. In order to compare it with Fig. 4(a) quantitatively, we used the time scale $T_{\text{real}} = 7$[sec]; the interval between two consecutive trades in the actual foreign exchange market is, on average, known [33] to be about 7[sec]. The spring constant in the actual market k is of the same magnitude as k_{micro}. It decays exponentially with the trade frequency f_{real}, which is also consistent with that of the microscopic model shown in Fig. 4(a). The real characteristic frequency in Fig. 4(b), however, is quite different from that of the microscopic model plotted in Fig. 4(a). This is an open problem.

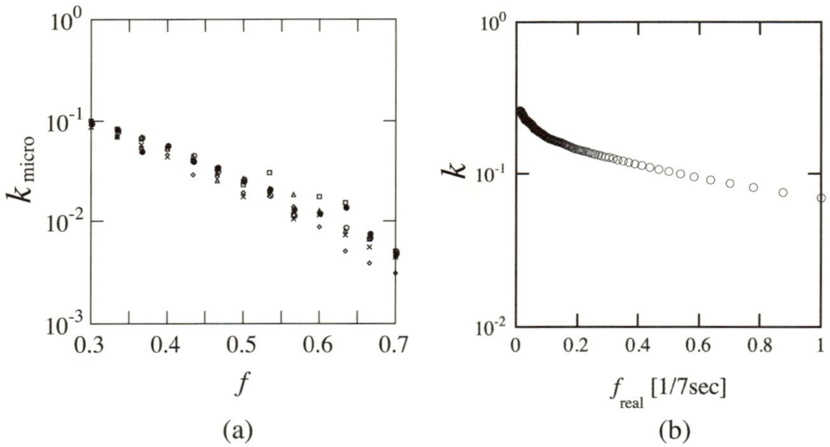

Fig. 4. (a) The scaling plot of the spring constant k_{micro} as a function of the trade frequency $f = N\alpha/3\Lambda$. The vertical axes are displayed in the logarithmic scale. The dealers' response to the price change for $c = 0.3$. We fix $\alpha = 0.0001, 0.001$ and 0.01 and change N (open circles, squares, and diamonds, respectively) and $N = 100, 1000$ and 10000 and change α (crosses, filled circles and triangles, respectively), while Λ is fixed to 1. Note that all points collapse onto a single curve. (b) The spring constant k defined in Eq. (6) as a function of the trade frequency scaled by the realistic time scale of the trades, $T_{\text{real}} = 7$[sec].

5 Summary

We first showed that triangular arbitrage opportunities exist in the foreign exchange market. The probability density function of the logarithm rate product ν has a sharp peak and fat tails. Once there is a triangular arbitrage opportunity, or if the logarithm rate product ν is positive, many traders will try to make profit through the triangular arbitrage transaction. This makes ν converge to a slightly negative value, thereby eliminating the opportunity. Triangular arbitrage opportunities nevertheless appear, because each rate r_x fluctuates strongly.

In Sect. 2, we introduced the macroscopic model, which contains the interaction caused by the triangular arbitrage transaction. We showed that the interaction is the reason of the sharp peak and the fat tails of the distribution of the logarithm rate product ν. In Sect. 3, we introduced the microscopic model, which consists of two systems of the ST model. The microscopic model reproduced the actual behavior of the logarithm rate product ν well. The microscopic model can describe more details than the macroscopic model, in particular, the skewness of the distribution of the logarithm rate product ν.

We finally explored in Sect. 4 the relation between the spring constant of the macroscopic model and the parameters in the microscopic model. The spring constant of the microscopic model k_{micro} can be scaled by the trade frequency f and it decays exponentially with f, which is consistent with the spring constant of the actual market k.

References

1. Mandelbrot BB (1963) J. Business 36:394–419
2. Fama EF (1970) J. Finance 25:383–423
3. Ding Z, Granger CWJ, Engle RF (1993) J. Empirical Finance 1:83–106
4. Dacorogna MM, Muller UA, Nagler RJ, Olsen RB, Pictet OV (1993) J. Int. Money Finance 12:413–438
5. Liu YH, Cizeau P, Meyer M, Peng CK, Stanley HE (1997) Physica A 245:437–440
6. Mantegna RN (1999) Eur. Phys. J. B 11:193–197
7. Plerou V, Gopikrishnan P, Rosenow B, Amaral LAN, Stanley HE (1999) Phys. Rev. Lett. 83:1471–1474
8. Plerou V, Gopikrishnan P, Rosenow B, Amaral LAN, Stanley HE (2000) Physica A 287:374–382
9. Kullmann L, Kertesz J, Mantegna RN (2000) Physica A 287:412–419
10. Mandelbrot BB (2001) Quant. Fin. 1:113–123
11. Mandelbrot BB (2001) Quant. Fin. 1:124–130
12. Kullmann L, Kertesz J, Kaski K (2002) Phys. Rev. E 66:026125
13. Onnela JP, Chakraborti A, Kaski K, Kertesz J, Kanto A (2003) Phys. Rev. E 68:056110
14. Mizuno T, Kurihara S, Takayasu M, Takayasu H (2004) In: Takayasu H (ed) The Application of Econophysics, Proceedings of the Second Nikkei Symposium, Springer-Verlag, Tokyo: pp. 24–29.
15. Tastan H (2006) Physica A 360:445–458
16. Toth B, Kertesz J (2006) Physica A 360:505–515
17. Jung WS, Chae S, Yang JS, Moon HT (2006) Physica A 361:263–271
18. Aiba Y, Hatano N, Takayasu H, Marumo K, Shimizu T (2002) Physica A 310:467–479
19. Aiba Y, Hatano N, Takayasu H, Marumo K, Shimizu T (2003) Physica A 324:253–257
20. Aiba Y, Hatano N, Takayasu H, Marumo K, Shimizu T (2004) In: Takayasu H (ed) The Application of Econophysics, Proceedings of the Second Nikkei Symposium. Springer-Verlag, Tokyo: pp. 18–23
21. Aiba Y, Hatano N (2004) Physica A 344:174–177
22. Aiba Y, Hatano N (2006) In: Chakrabarti BK, Chakraborti A, Chatterjee A (Eds) Trends and Perspectives in Econophysics and Sociophysics. Wiley-VCH, Berlin, to appear
23. Aiba Y, Hatano N (2006) Physica A, submitted
24. Mavirides M (1992) Quorum Books, New York
25. Moosa I (2001) Quantitative Finance 1:387–390
26. Mantegna RN, Stanley HE (1999) Cambridge University Press, Cambridge: pp. 64–67
27. Engle RF (1982) Econometrica 50:987–1002
28. Bollerslev T (1986) J. Econometrics 31:307–327

29. Bouchaud JP, Potters M (2000) Cambridge University Press, Cambridge: p. 34 and pp. 56–63
30. Takayasu H, Takayasu M, Okazaki MP, Marumo K, Shimizu T (2000) In: Novak MM (ed) Paradigms of Complexity: Fractals and Structures in the Sciences, World Scientific, Singapore: pp. 243–258
31. Mantegna RN (1994) Phys. Rev. E 49:4677–4683
32. Sato AH, Takayasu H (1998) Physica A 250:231–252
33. Ohira T, Sazuka N, Marumo K, Shimizu T, Takayasu M, Takayasu H (2002) Physica A 308:368–374

Modelling Limit Order Financial Markets

Robin Stinchcombe

Rudolf Peierls Centre for Theoretical Physics, Oxford University, 1 Keble Road, OXFORD, OX1 3NP, UK. stinch@thphys.ox.ac.uk

Summary. Financial markets are collective systems with out of equilibrium stochastic dynamics. Of these, the simplest is the limit order market, where an order book records placement and removal of orders to buy or sell, and their settlement. Such systems can be modelled by regarding the orders as depositing, evaporating or annihilating, at prescribed rates, on a price axis. The construction and analysis of such models and their properties and limitations will be discussed here, starting from available electronic temporal limit order market data. This makes it possible to infer the stochastic dynamic processes which operate in real markets and how their rates are connected between themselves and to the market condition. Analytic analysis for steady state profiles and price evolution of such models will be outlined. A concluding discussion reviews the work and raises further issues.

1 Introduction

In limit order markets [1], orders to buy ("bids") or sell ("asks") (for a particular stock) are recorded in an order book, which evolves until transactions eventually occur.

The price (x) axis is fundamental in modelling such markets, as is the movement of the limiting orders to buy (ie the highest or "best" bid) or sell (the lowest ask) and of the spread (the gap between them) (Fig. 1).

The market evolution is tied in with the specific order processes (placement, etc) and how their probabilities depend on price, etc.

The condition of the market will affect the action of traders, and this introduces feedback effects which could strongly modify the evolution of the market [2,3], (through herding for example, [4]).

In common with crowds, traffic, etc., markets are stochastic collective nonequilibrium systems, so their understanding benefits from approaches like those used for corresponding physical systems [5,6].

Such approaches are of the zero-intelligence type (cf [7]), and earlier work along such lines for limit order markets includes that of [8–11], which differ in the processes incorporated (diffusion versus placement, etc).

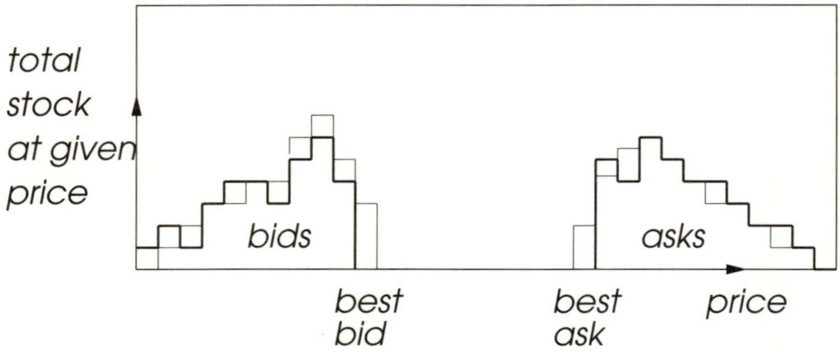

Fig. 1. Successive instantaneous states of limit order market/book given as accumulated bids and asks (for a particular stock) as a function of price.

Market data can in principle identify the processes and decide which type of model is appropriate. Indeed, the order book allows a detailed quantitative assessment of the market condition to be made at any time, and (from its update) the actual processes to be identified. This has been exploited in [12]. This latter work is followed closely in the next section (Sect. 2).

A resulting minimal limit order market model and some of its properties [12] (see also [13–15]) are given in section Sect. 3 while Sect. 4 introduces some generalisations and a basic analytical approach to the model, somewhat analogous to previous work [16] for a forerunner model [10]. Resulting analyses of steady state and dynamic behaviour, outlined in Sect. 5, have some contact with perspectives in [17,18] and with further approximate analytical work [19]. Sec. 6 is a concluding discussion.

2 Brief review of previous data analysis identifying processes and generic features of limit order markets

We here briefly review the empirical evidence [12] which points to a nonequilibrium model with the processes of deposition, transaction and cancellation of orders, with probabilities dependent in a specific way on price and market condition.

The data analysed was from the ECN part of NASDAQ, specifically from four stocks including Cisco (CSCO) and Dell (DELL). Both static properties (cf earlier studies) and also dynamic ones were investigated.

2.1 Steady state data

Clustering was evident in both the order size and order price histograms, while order lifetimes showed power-law distributions. Aggregated order size profiles on the price axis showed convexity and strong peaking at the gap edges (Fig. 2), consistent with overdiffusive price behaviour. Non-zero, time-dependent virtual impact functions were seen[1].

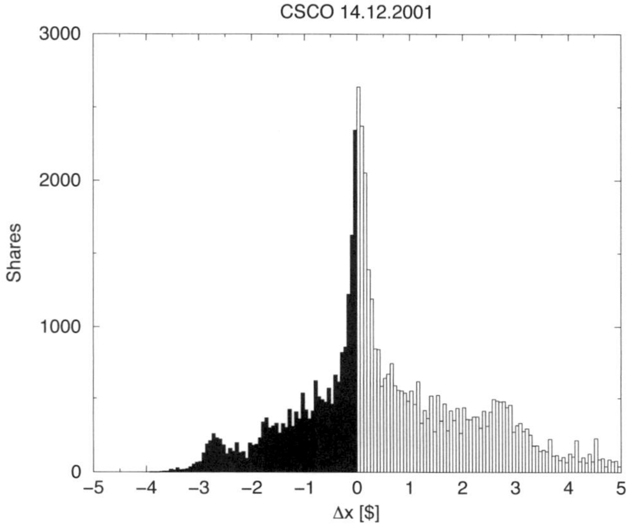

Fig. 2. Histogram of aggregated orders of CSCO versus the price measured from the relevant gap edge. The left and right distributions are the buy and sell side, respectively.

2.2 Dynamic data

The processes identified were placement (deposition, characteristic rate δ_X), or withdrawal (evaporation, rate η_X), of orders of either type $X = A, B$ (ask, bid); and transaction (AB pair annihilation, rate α).

The event frequencies were seen to depend on spatial position Δx (price relative to gap edges) since, as could be expected, the activity is concentrated near the gap edges (Fig. 3) (cf the order price distributions Fig. 2).

Rates were seen to be fluctuating, but correlated (with δ_X/η_X roughly constant) in time (Fig. 4). Deposition events dominate, with evaporation next most frequent, followed by annihilation. The degree of diffusion was seen to be extremely small. Algebraically-time-dependent rate auto- and cross-correlation functions were found.

Feedback effects were seen in the dependence of process frequencies on time-dependent characteristics of the market: in particular the spatial width of deposition, etc, (Fig. 3) depends roughly linearly on the gap $g(t) = a(t) - b(t)$; here the best bid and best ask at time t are denoted by $b(t), a(t)$.

[1] For further details, see [12].

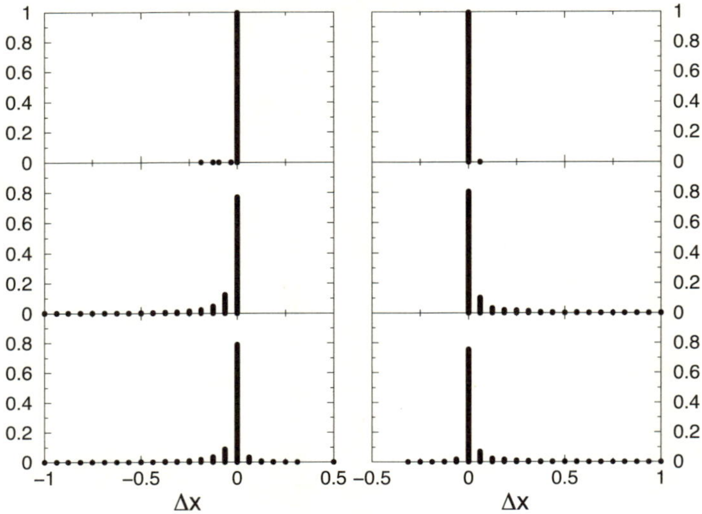

Fig. 3. Frequency of event versus the price to the relevant edge. Left (right) column refers to bid (ask) side. The events in each column, are, from the top, annihilation, evaporation, deposition.

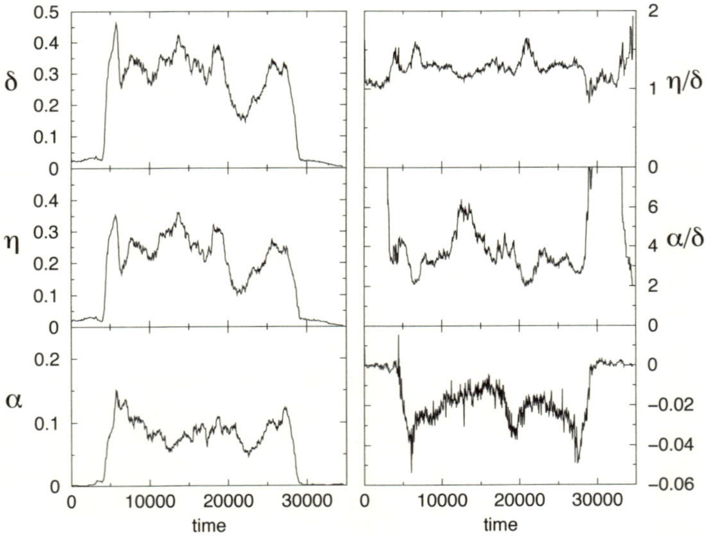

Fig. 4. On the left, from top to bottom: bid rates of deposition δ, evaporation η and annihilation α in events/second, time averaged over a window of 1500 seconds of a day of DELL. Right: ratios δ/η and δ/α, and $\delta - \eta - \alpha$

3 Resulting minimal model and properties

3.1 Two-species limit order market model with feedback

The empirical evidence outlined in the preceding section indicates a collective model [12] with two species of particle (bids and asks) containing the processes of deposition, evaporation and annihilation. Diffusion appears to be negligible. Moreover, the evidence suggests that the associated probabilities should depend on price relative to the extremum and on the market condition, through the gap size. The price dependence allows annihilation (a consequence of depositing a 'market order' immediately transacted at the appropriate edge) to be included in the deposition process. A minimal form of the model [12] takes all orders to be of the same size ('particles'), all events to be independent and time independent, and allows at most one order to be deposited in each time step.

The events are then:

(i) with probability δ_A an A particle (ask) is deposited at site x drawn at random from a probability density function $P_A(x,t)$; whenever x lies in the bulk of the B distribution, the annihilation is regarded as occurring at the relevant gap edge. Similarly for the B particle deposition.

(ii) particle A (or B) can evaporate with probability η_A (or η_B) at site x drawn from the appropriate $P(x,t)$.

A basic form incorporating the observed feedback takes each $P(x,t)$ to be centred on the relevant gap edge and to be Gaussian, with gap-dependent variance of the form

$$\sigma(t) = Kg(t) + C. \tag{1}$$

A set δ, η, K, C for each species comprise the parameters of the model. It can be convenient [13], [19] to work with dimensionless variables (scaled by characteristic measures of time, price interval, and share number).

Simulation results from this simplest version of the model are outlined in the next subsection. In Sect. 4 various generalisations including order size dependences and non-Gaussian P's will be considered.

3.2 Results from the model

Monte Carlo simulations were carried out for a version of the minimal model in which the parameters δ, η, K, C are taken to be species-independent (the same for bids and asks). The model gives

(i) power law tails of returns

(ii) volatility clustering

(iii) rich market evolution

(iv) asymptotically diffusive Hurst behaviour (Hurst exponent increasing with time from 1/4 (underdiffusive) to 1/2 (diffusive)).

The volatility clustering occurs in various parameter ranges and is connected with the power law decay of the gap autocorrelation function (compare [10]).

The market evolution given by the model (simulation) (Fig. 5) is strongly dependent on the parameter K measuring the feedback (via the gap) from the market condition. It can mimic behaviours ranging from gradual inflation to crashes. Analytic work on generalisations of the model will be outlined in the next two sections.

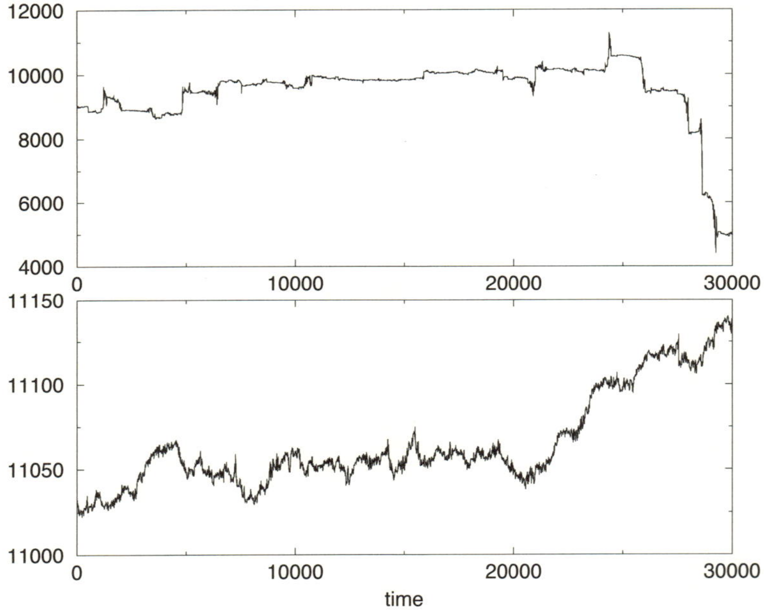

Fig. 5. Two time series of the market model with parameters $\alpha = 1/2$, $\delta = 1/100$, $C = 3$, and $K = 2$ (upper graph) and $K = 1$ (lower graph), showing the dramatic influence of K on the system's evolution.

4 Generalisations; basic analytic approach

4.1 Generalisations of the model

A more complete description of the market is provided by the sizes m of each of the orders (of either type, bid or ask) at a given position x on the price axis.

The instantaneous state of the order book then corresponds to a column at each such position, each of total (non-negative) height corresponding to the sum of the m's for the orders at that price. At any time t between transactions the bid and ask columns are in different regions on the price axis, ie in the two 'bands' $x \leq b(t)$ and $x \geq a(t)$ respectively (Fig. 6).

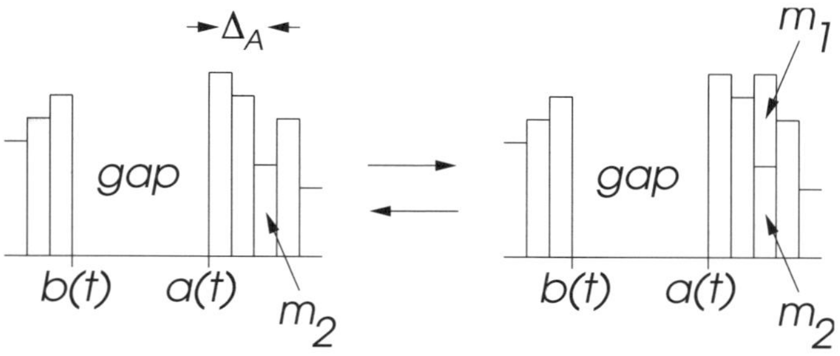

Fig. 6. Deposition (at rate $\delta_X P_{X,\delta}(\Delta_A/\sigma, m_1, m_2)$) and evaporation (at rate $\eta_X P_{X,\eta}(\Delta_A/\sigma, m_1, m_2)$) of ask order of size m_1 in ask band at Δ_A from band edge.

Then, from the analysis of the data, the limit order market processes ν are closely represented as: deposition ($\nu = \delta$, now incorporating possible consequent transaction), and evaporation ($\nu = \eta$), at any place x on the on the price axis, of orders of two species, $X = A, B$ (asks or bids), of any size m. The probabilities $P_{X,\nu}$ depend on (a) the place Δ_X on the price axis where the process occurs, relative to the appropriate gap edge, and (b) the market condition, through the ratio $r_X(x,t) \equiv \Delta_X/[C+Kg(t)]$ (where $\Delta_A = x-a(t)$, $\Delta_B = x - b(t)$), and also on the order sizes involved. Specifically (suppressing the dependence on $r_X(x,t)$), $P_{X,\nu}(m_1, m_2)$ denotes probabilities for deposition of m_1 onto m_2 at x ($\nu = \delta$) and evaporation of m_1 to leave m_2 at x ($\nu = \eta$) (illustrated for asks in Fig. 6). (Using normalised probabilities for each species and each process $\nu = \delta, \eta$ there also remains an overall rate ν_X.)

4.2 Basic approach to generalised model

A master equation can then be written, in terms of these probabilities, for the evolution of the probability P_C of the market being in any specified state C. The equation is intractable as it stands, but some progress can be made by employing a "mean field factorisation" in which configurational probabilities are approximated by products of column probabilities. The particular way this is done corresponds to the 'independent interval approximation' in related exclusion models [20, 21], and a previous application of the approximation to the limit order market has been given in [19].

The basic quantity then to be determined is the probability $P_X(m; x, t)$ of a column of height m of either species $X = A, B$ at each price x and time t. This satisfies, for x inside the corresponding 'band',

$$\partial P_X(m;x,t)/\partial t =$$
$$\sum_{m'} \Big(\delta_X[P_{X,\delta}(m',m-m')P_X(m-m';x,t) - P_{X,\delta}(m',m)P_X(m;x,t)]$$
$$+\,\eta_X[P_{X,\eta}(m',m)P_X(m+m';x,t) - P_{X,\eta}(m',m-m')P_X(m;x,t)]\Big) \quad (2)$$

Processes at other x's (at or beyond the corresponding band edge) typically cause the band edge to move, resulting in self-consistent equations for the evolution of the best bid and ask. Such aspects are easier to describe for the 'exclusion' version of the model (each m restricted to 0,1, cf Sect. 3), and we continue with that case.

5 Simplified model analysis

For the exclusion model, the quantity of interest is the (coarse-grained) average m for each X, x, t, ie the density profile $\rho_X(x,t)$. For the illustrative case of asks it satisfies the following mean field equation

$$\partial \rho_A(x,t)/\partial t = \delta_A P_{A\delta}(r_A(x,t))(1-\rho_A(x,t)) - \eta_A P_{A\eta}(r_A(x,t))\rho_A(x,t)$$
$$-\,\delta(x-a(t))\rho_A(x,t)\int_{a(t)}^{\infty} \delta_B P_{B\delta}(r_B(x',t))\,dx' \quad (3)$$

and similarly for bids. Here we do not assume the $P_{X\nu_X}$ to be Gaussian functions of r_X. The integral term contains the effects of market orders transacted at the band edges (see Subsect. 5.2).

5.1 Steady state bulk bid and ask profiles

Fig. 4 suggests that the ratio $\gamma_X \equiv \eta_X P_{X\eta}(r_X(x,t))/\delta_X P_{X\delta}(r_X(x,t))$ is approximately time independent. That is consistent with an adiabatically developing solution of (3) of the form $\rho_A \sim 1/(1+\gamma_A)$ for $x > a$, and similarly for bids.

From data used to obtain Fig. 3, $P_{A\delta}$ actually falls off with increasing positive Δ_A faster than $P_{A\eta}$, which makes the ask density profile $\rho_A(x)$ a rapidly decreasing function of Δ_A. Similarly for the bids. This mean field argument gives a qualitative account of the schematic shape of the order price distributions (density profiles) Fig. 2 inside the bid and ask regions.

5.2 Evolution of prices and spread

The evolution of the gap edges can be investigated by inserting a cut-off profile ($\rho_A = f(x,t)\Theta(x-a(t))$) into (3), and correspondingly for bids. Matching the most singular (delta-function) terms arising then gives two coupled differential equations for $a(t), b(t)$ in terms of integrals of the form $\int_{g(t)}^{\infty} P_{X\delta}(r')\,dr'$.

Explicit forms can then be obtained for the evolution of the best bid and ask (and hence the spread) for particular forms of the $P_{X\delta}$, capturing some elements of trends. Note that the approach includes the feedback from the market condition and the results show a strong dependence on the parameter K.

However, the resulting description is deterministic - this is due to the mean field factorisation. Adding noise terms to the equations restores the proper stochastic character, and the resulting description makes contact with previous generalised random walk analyses of price evolution in the model [17, 18].

Procedures similar to those just illustrated can be applied to the generalised model with various forms of the rate distributions $P_{X,\nu}(r_X, m_1, m_2)$. Analyses are much simplified when approximate symmetries exist between bid and ask regions.

6 Concluding discussion

While the limit order market data clearly supports nonequilibrium two-species models with processes and market feedback of the type here discussed, there remains some flexibility, eg between a minimal version with orders of one size, or a generalised one, and between alternative functional forms of the $P_{X,\nu}(r_X, ...)$. The simulation and analytic work suggest that while some features (including power law tails of returns, volatility clustering, and market evolution, which is set largely by the strength of the feedback) are robust, others (such as density profiles near the edges) are affected by such flexibility.

The main point to be stressed however is that the forms of the $P_{X,\nu}$ and especially their argument r_X, which together encapsulate the market feedback, are purely empirical. What is presently missing is an account of how traders' perception of the market condition leads to such a form of feedback. Whether such approaches as game theory (eg the minority game, [22, 23]) will provide this remains unclear. It would be nice if the explanations turn out to be universal, as could be the case if for example markets are near critical (as suggested in some recent work, [24]).

Acknowledgements

I am most grateful to Bikas Chakrabarti and Arnab Chatterjee for their invitation. I wish to thank Damien Challet for his generosity in the collaborations on which much of this work is based. The work was supported by EPSRC under the Oxford Condensed Matter Theory Grants GR/R83712/01 and GR/M04426/01.

References

1. See e.g. Varian HR (2002) Intermediate Microeconomics. 6 ed, Norton, New York
2. Bouchaud JP, Potters M (2000) Theory of Financial Risks, Cambridge University Press, Cambridge
3. Mantegna RN, Stanley HE (2000) Introduction to Econophysics, Cambridge University Press, Cambridge
4. Lux T, Marchesi M (1999) Nature 397:498-500
5. van Kampen NG (1992) Stochastic Processes in Physics and Chemistry, Elsevier, Amsterdam
6. See e.g. Stinchcombe RB (2001) Advances in Physics 50:431, and references therein
7. Gode D, Sunder S (1993) J. of Political Economy 101:119
8. Bak P, Shubik M, Pakzuski M (1997) Physica A 246:430
9. Eliezer D, Kogan II (1999) Capital Markets Abstracts, Market Microstructure 2:3; Chan DLC, Eliezer D, Kogan II (2001) eprint cond-mat/0101474
10. Maslov S (2000) Physica A 278:571
11. Maslov S, Mills M (2000) Physica A 299:234
12. Challet D, Stinchcombe R (2001) Physica A 300:285
13. Daniels MG, Farmer JD, Guillemot L, Iori G, Smith E (2003) Phys. Rev. Lett. 90:108102
14. Challet D, Stinchcombe R (2003) Physica A 324:141
15. Bouchaud J-P, Mézard M, Potters M (2002) Quant. Fin. 2:251
16. Slanina F (2001) Phy. Rev. E 64:056136
17. Challet D, Stinchcombe R (2003) Quant. Fin. 3:165
18. Willmann RD, Schütz GM, Challet D (2002) Physica A 316:526
19. Smith E, Farmer JD, Gillemot L, Krishnamurthy S (2003) Quant. Fin. 3:481
20. Majumdar SN, Krishnamurthy S, Barma M (2000) J. Stat. Phys. 99:1
21. Reis F, Stinchcombe R (2004) Phys. Rev. E 70:036109; (2005) Phys. Rev. E 71:026110; (2005) Phys. Rev. E 72:031109
22. Challet D, Zhang Y-C (1997) Physica A 246:407
23. Challet D, http://www.unifr.ch/econophysics/minority
24. Raffaelli G, Marsili M, e-print physics/0508159

Two Fractal Overlap Time Series and Anticipation of Market Crashes

Bikas K. Chakrabarti[1], Arnab Chatterjee[1] and Pratip Bhattacharyya[1,2]

[1] Theoretical Condensed Matter Physics Division and Centre for Applied Mathematics and Computational Science, Saha Institute of Nuclear Physics, 1/AF Bidhannagar, Kolkata-700064, India.
bikask.chakrabarti@saha.ac.in, arnab.chatterjee@saha.ac.in
[2] Physics Department, Gurudas College, Narkeldanga, Kolkata 700 054, India
pratip.bhattacharyya@saha.ac.in

Summary. We find prominent similarities in the features of the time series for the (model earthquakes or) overlap of two Cantor sets when one set moves with uniform relative velocity over the other and time series of stock prices. An anticipation method for some of the crashes have been proposed here, based on these observations.

1 Introduction

Capturing dynamical patterns of stock prices are major challenges both for epistemologists as well as for financial analysts [1]. The statistical properties of their (time) variations or fluctuations [1] are now well studied and characterized (with established fractal properties), but are not very useful for studying and anticipating their dynamics in the market. Noting that a single fractal gives essentially a time averaged picture, a minimal two-fractal overlap time series model was introduced [2–4] to capture the time series of earthquake magnitudes. We find that the same model can be used to mimic and study the essential features of the time series of stock prices.

2 The two fractal-overlap model of earthquake

Let us consider first a geometric model [2–5] of the fault dynamics occurring in overlapping tectonic plates that form the earth's lithosphere. A geological fault is created by a fracture in the earth's rock layers followed by a displacement of one part relative to the other. The two surfaces of the fault are known to be self-similar fractals. In the model considered here [2–5], a fault is represented by a pair of overlapping identical fractals and the fault dynamics arising out of

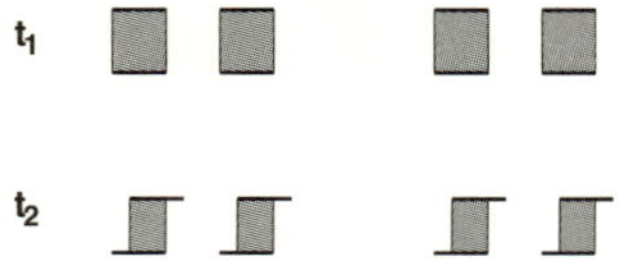

Fig. 1. The overlap of two identical Cantor sets of dimension $\ln 2/\ln 3$ at generation $n = 2$ as one moves over the other with uniform velocity. The total measure O of the overlap (total shaded region) varies with time and are shown for two different time instances.

the relative motion of the associated tectonic plates is represented by sliding one of the fractals over the other; the overlap O between the two fractals represents the energy released in an earthquake whereas $\log O$ represents the magnitude of the earthquake. In the simplest form of the model each of the two identical fractals is represented by a regular Cantor set of fractal dimension $\log 2/\log 3$ (see Fig. 1). This is the only exactly solvable model for earthquakes known so far. The exact analysis of this model [5] for a finite generation n of the Cantor sets with periodic boundary conditions showed that the probability of the overlap O, which assumes the values $O = 2^{n-k} (k = 0, \ldots, n)$, follows the binomial distribution F of $\log_2 O = n - k$ [6]:

$$\begin{aligned} \Pr\left(O = 2^{n-k}\right) &\equiv \Pr\left(\log_2 O = n - k\right) \\ &= \binom{n}{n-k} \left(\frac{1}{3}\right)^{n-k} \left(\frac{2}{3}\right)^{k} \equiv F(n-k). \end{aligned} \tag{1}$$

Since the index of the central term (i.e., the term for the most probable event) of the above distribution is $n/3 + \delta$, $-2/3 < \delta < 1/3$, for large values of n Eq. (1) may be written as

$$F\left(\frac{n}{3} \pm r\right) \approx \binom{n}{n \pm r} \left(\frac{1}{3}\right)^{\frac{n}{3} \pm r} \left(\frac{2}{3}\right)^{\frac{2n}{3} \mp r} \tag{2}$$

by replacing $n - k$ with $n/3 \pm r$. For $r \ll n$, we can write the normal approximation to the above binomial distribution as

$$F\left(\frac{n}{3} \pm r\right) \sim \frac{3}{\sqrt{2\pi n}} \exp\left(-\frac{9r^2}{2n}\right) \tag{3}$$

Since $\log_2 O = n - k = \frac{n}{3} \pm r$, we have

$$F\left(\log_2 O\right) \sim \frac{1}{\sqrt{n}} \exp\left[-\frac{\left(\log_2 O\right)^2}{n}\right], \tag{4}$$

not mentioning the factors that do not depend on O. Now

$$F\left(\log_2 O\right) \mathrm{d}\left(\log_2 O\right) \equiv G(O) \mathrm{d}O \tag{5}$$

where

$$G(O) \sim \frac{1}{O} \exp\left[-\frac{(\log_2 O)^2}{n}\right] \tag{6}$$

is the log-normal distribution of O. As the generation index $n \to \infty$, the normal factor spreads indefinitely (since its width is proportional to $\sqrt{n}$) and becomes a very weak function of O so that it may be considered to be almost constant; thus $G(O)$ asymptotically assumes the form of a simple power law with an exponent that is independent of the fractal dimension of the overlapping Cantor sets [6]:

$$G(O) \sim \frac{1}{O} \text{ for } n \to \infty. \tag{7}$$

3 The Cantor set overlap time series

We now consider the time series $O(t)$ of the overlap set (of two identical fractals [4,5]), as one slides over the other with uniform velocity. Let us again consider two regular cantor sets at finite generation n. As one set slides over the other, the overlap set changes. The total overlap $O(t)$ at any instant t changes with time (see Fig. 2(a)). In Fig. 2(b) we show the behavior of the

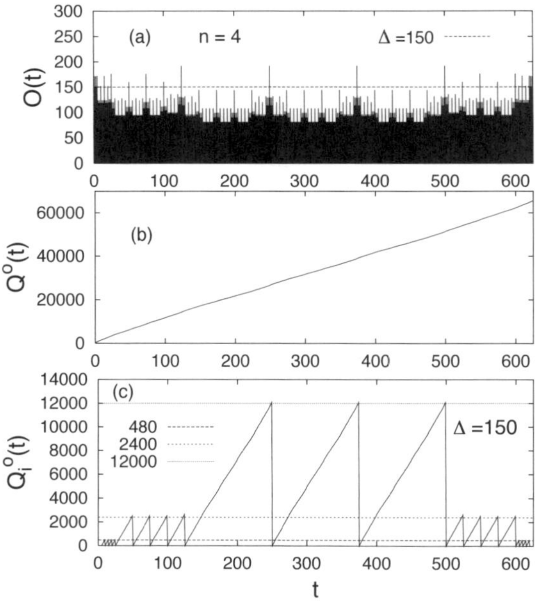

Fig. 2. (a) The time series data of overlap size $O(t)$ for a regular Cantor set of dimension ln4/ln5 at generation $n = 4$. (b) Cumulative overlap $Q^o(t)$ and (c) the variation of the cumulative overlap $Q_i^o(t)$ for the same series, where Q is reset to zero after any big event of size greater than $\Delta = 150$.

cumulative overlap [4] $Q^o(t) = \int_0^t O(\tilde{t})d\tilde{t}$. This curve, for sets with generation $n = 4$, is approximately a straight line [4] with slope $(16/5)^4$. In general, this curve approaches a strict straight line in the limit $a \to \infty$, asymptotically, where the overlap set comes from the Cantor sets formed of $a - 1$ blocks, taking away the central block, giving dimension of the Cantor sets equal to $\ln(a-1)/\ln a$. The cumulative curve is then almost a straight line and has then a slope $\left[(a-1)^2/a\right]^n$ for sets of generation n. If one defines a 'crash' occurring at time t_i when $O(t_i) - O(t_{i+1}) \geq \Delta$ (a preassigned large value) and one redefines the zero of the scale at each t_i, then the behavior of the cumulative overlap $Q_i^o(t) = \int_{t_{i-1}}^t O(\tilde{t})d\tilde{t}$, $\tilde{t} \leq t_i$, has got the peak value 'quantization' as shown in Fig. 2(c). The reason is obvious. This justifies the simple thumb rule: one can simply count the cumulative $Q_i^o(t)$ of the overlaps since the last 'crash' or 'shock' at t_{i-1} and if the value exceeds the minimum value (q_o), one can safely extrapolate linearly and expect growth upto αq_o here and face a 'crash' or overlap greater than Δ ($= 150$ in Fig. 2). If nothing happens there, one can again wait upto a time until which the cumulative grows upto $\alpha^2 q_o$ and feel a 'crash' and so on ($\alpha = 5$ in the set considered in Fig. 2).

4 The stock price time series

We now consider some typical stock price time-series data, available in the internet. The data analyzed here are for the New York Stock Exchange (NYSE) Indices [7]. In Fig. 3(a), we show that the daily stock price $S(t)$ variations for about 10 years (daily closing price of the 'industrial index') from January 1966 to December 1979 (3505 trading days). The cumulative $Q^s(t) = \int_0^t S(t)dt$ has again a straight line variation with time t (Fig. 3(b)). Similar to the Cantor set analogy, we then define the major shock by identifying those variations when $\delta S(t)$ of the prices in successive days exceeded a preassigned value Δ (Fig. 3(c)). The variation of $Q_i^s(t) = \int_{t_{i-1}}^{t_i} S(\tilde{t})d\tilde{t}$ where t_i are the times when $\delta S(t_i) \leq -1$ show similar geometric series like peak values (see Fig. 3(d)); see [8].

We observed striking similarity between the 'crash' patterns in the Cantor set overlap model and that derived from the data set of the stock market index. For both cases, the magnitude of crashes follow a similar pattern — the crahes occur in a geometric series.

A simple 'anticipation strategy' for some of the crashes may be as follows: If the cumulative $Q_i^s(t)$ since the last crash has grown beyond $q_0 \simeq 8000$ here, wait until it grows (linearly with time) until about 17,500 ($\simeq 2.2q_0$) and expect a crash there. If nothing happens, then wait until $Q_i^s(t)$ grows (again linearly with time) to a value of the order of 39,000 ($\simeq (2.2)^2 q_0$) and expect a crash, and so on.

The same kind of analysis for the NYSE 'utility index', for the same period, is shown in Figs. 4.

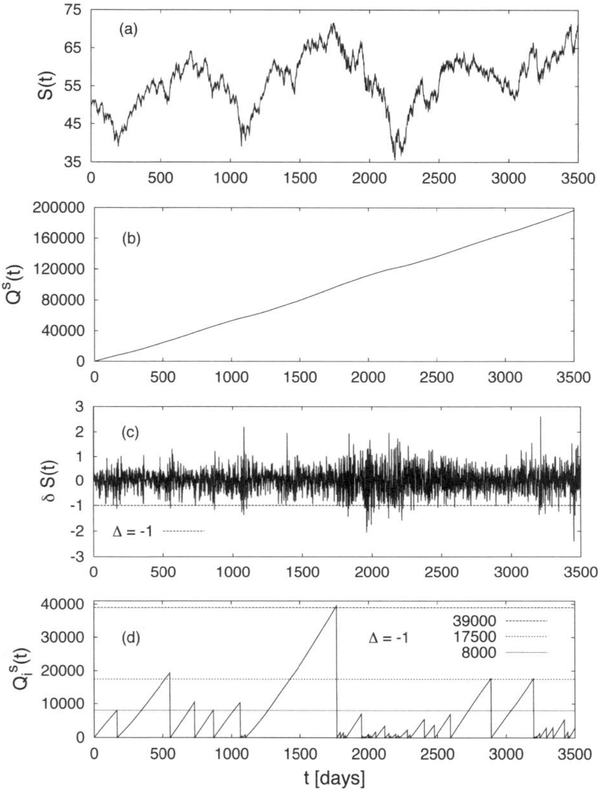

Fig. 3. Data from New York Stock Exchange from January 1966 to December 1979: industrial index [7]: (a) Daily closing index $S(t)$ (b) integrated $Q^s(t)$, (c) daily changes $\delta S(t)$ of the index $S(t)$ defined as $\delta S(t) = S(t+1) - S(t)$, and (d) behavior of $Q_i^s(t)$ where $\delta S(t_i) > \Delta$. Here, $\Delta = -1.0$ as shown in (c) by the dotted line (from [8]).

5 Summary

Based on the formal similarity between the two-fractal overlap model of earthquake time series and of the stock market, we considered here a detailed comparison. We find, the features of the time series for the overlap of two Cantor sets when one set moves with uniform relative velocity over the other looks somewhat similar to the time series of stock prices. We analyze both and explore the possibilities of anticipating a large (change in Cantor set) overlap or a large change in stock price. An anticipation method for some of the crashes has been proposed here, based on these observations.

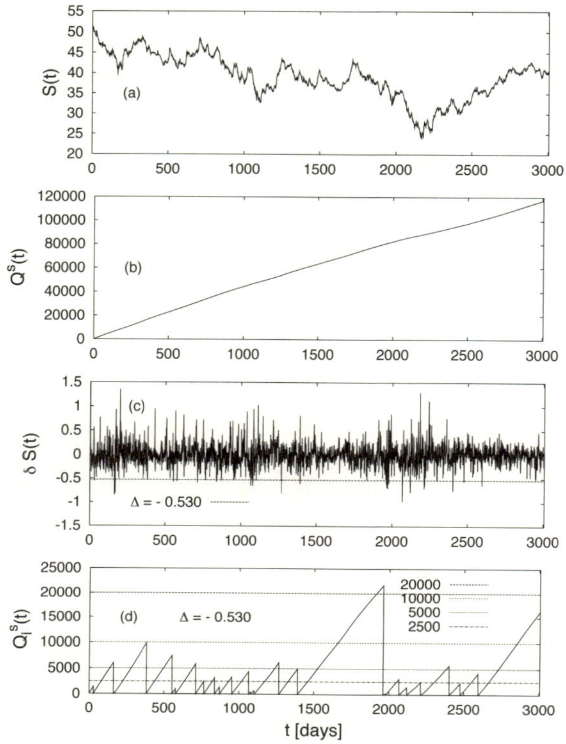

Fig. 4. Data from New York Stock Exchange from January 1966 to December 1979: utility index [7]: (a) Daily closing index $S(t)$ (b) integrated $Q^s(t)$, (c) daily changes $\delta S(t)$ of the index $S(t)$ defined as $\delta S(t) = S(t+1) - S(t)$, and (d) behavior of $Q_i^s(t)$ where $\delta S(t_i) > \Delta$. Here, $\Delta = -0.530$ as shown in (c) by the dotted line.

References

1. Sornette D (2003) Why Stock Markets Crash? Princeton Univ. Press, Princeton; Mantegna RN, Stanley HE (1999) Introduction to Econophysics. Cambridge Univ. Press, Cambridge
2. Chakrabarti BK, Stinchcombe RB (1999) Physica A 270:27-34
3. Pradhan S, Chakrabarti BK, Ray P, Dey MK (2003) Phys. Scr. T106:77-81
4. Pradhan S, Chaudhuri P, Chakrabarti BK (2004) in Continuum Models and Discrete Systems, Ed. Bergman DJ, Inan E, Nato Sc. Series, Kluwer Academic Publishers, Dordrecht, pp.245-250; cond-mat/0307735
5. Bhattacharyya P (2005) Physica A 348:199-215
6. Bhattacharyya P, Chatterjee A, Chakrabarti BK (2006) physics/0512036; Physica A, to be published
7. NYSE Daily Index Closes from http://www.unifr.ch/econophysics/
8. Chakrabarti BK, Chatterjee A, Bhattacharyya P (2006) in Takayasu H (Ed) Practical Fruits of Econophysics, Springer, Tokyo, pp. 107-110.

The Apparent Madness of Crowds: Irrational Collective Behavior Emerging from Interactions among Rational Agents

Sitabhra Sinha

The Institute of Mathematical Sciences, C. I. T. Campus, Taramani,
Chennai - 600 113, India. sitabhra@imsc.res.in

Standard economic theory assumes that agents in markets behave rationally. However, the observation of extremely large fluctuations in the price of financial assets that are not correlated to changes in their fundamental value, as well as the extreme instance of financial bubbles and crashes, imply that markets (at least occasionally) do display irrational behavior. In this paper, we briefly outline our recent work demonstrating that a market with interacting agents having bounded rationality can display price fluctuations that are *quantitatively* similar to those seen in real markets.

1 Introduction

It has long been debated in the economic literature whether markets exhibit irrational behavior [1]. The historical observations of apparent financial "bubbles", in which the demand (and therefore, the price) for certain assets rises to unreasonably high levels within a very short time only to come crashing down later [2], imply that markets act irrationally, because the rapid price changes are not associated with changes in the fundamental value of the assets. Believers in rational expectation theory argue that the price rise actually reflects the market's expectations about the long-term prospect of these assets and the large fluctuations are just rapid adjustments of these expectations in the light of new information [3]. These advocates of the "efficient market" school of thought claim that popular descriptions of speculative mania (e.g., in Ref. [4]) have been often exaggerated. However, critics point out that the market's estimate of the long-term value of an asset is a quantity that cannot be measured, and therefore, it is difficult to verify whether historical bubbles were indeed rational outcomes.

In this paper, we take an intermediate position between these two opposing camps. We assume that individual agents do behave in a rational manner, where rationality is identified with actions conducive to market equilibrium.

In other words, rational agents will act in such a way that the market is "balanced", exhibiting neither excess demand nor supply. Therefore, we expect only small fluctuations about the equilibrium when we have a large ensemble of non-interacting agents. In the model presented in this paper, market behavior is described by the collective decision of many interacting agents, each of whom choose whether to buy or sell an asset based on the limited information available to them about its prospects. In isolation, each agent behaves so as to drive the market to equilibrium. We investigate the possibility that interactions between such agents can severely destabilize the market equilibrium. In fact, we show that when agents are allowed to modify their interactions with neighbours, based on information about their past performance in the market, this results in the market becoming unbalanced and exhibiting extremely large fluctuations that are quantitatively similar to those seen in real markets.

2 Collective irrationality in an agent-based model

In this section, we present an agent-based model of the fluctuation of demand for a particular asset. The agents are assumed to be operating under bounded rationality, i.e., they try to choose between buying and selling the asset based on information about the action of their immediate neighbors and how successful their previous choices were. The fundamental value of the asset is assumed to be unchanged throughout the period. From the "efficient markets" hypothesis, we should therefore expect to see only small departures from the equilibrium. In addition, the agents are assumed to have limited resources, so that they cannot continue to buy or sell indefinitely. However, instead of introducing explicit budget constraints [5], we have implemented gradually diminishing returns for a decision that is taken repeatedly.

We assume that all agents are placed on a lattice, each site being occupied by one agent. An agent can only interact with its immediate neighbors on the lattice. In the simulations reported here, we have considered a two-dimensional hexagonal lattice, so that the number of neighbors is $z = 6$. At any given time t, the state of an agent i is fully described by two variables: its choice, S_i^t, and its belief about the outcome of the choice, θ_i^t. The choice can be either *buy* ($= +1$) or *sell* ($= -1$), while the belief can vary continuously over a range. The behavior of the agent over time can then be described by the equations governing the dynamics of S and θ,

$$S_i^{t+1} = \text{sign}(\Sigma_j J_{ij}^t S_j^t - \theta_i^t), \;\; \theta_i^{t+1} = \theta_i^t + \mu_i S_i^{t+1}, \tag{1}$$

where, J_{ij}^t measures the degree of interaction between neighboring agents. The adaptation rate, μ_i, governs the time-scale of diminishing returns, over which the agent switches from one choice to another in the absence of any interactions between agents. The overall state of the market at any given time is described by the fractional excess demand, $M^t = (1/N)\Sigma_j S_j^t$.

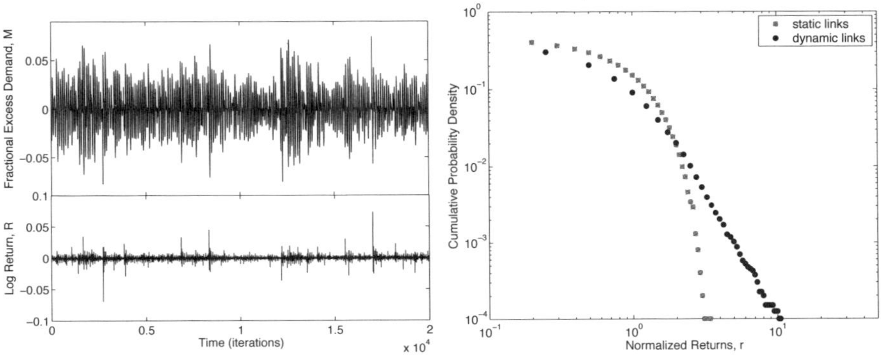

Fig. 1. Spontaneous fluctuations through dynamic interactions in a system of 400×400 agents. The adaptation rates, μ_i, are chosen from an uniform random distribution between $[0, 0.1]$. (Left) The time series of the fractional excess demand, M (top) and the corresponding logarithmic return, R (bottom), exhibiting significant number of large fluctuations. (Right) The cumulative probability density function of normalized returns, r, comparing the case where the J_{ij} are fixed in time (squares) with the situation when the J_{ij} change over time (circles). The latter shows a power-law decay with an exponent close to 3.

In previous work [6], we have shown that, if the interactions between agents do not change over time (i.e., $J^t_{ij} = J$, a constant), then M shows only small fluctuations about 0. This accords with the "efficient markets" hypothesis that any transient imbalance in the demand or supply of the asset is quickly corrected through the appropriate response of agents, so that the market remains more or less in equilibrium. However, if the agents have access to global information about the market (i.e., M), under certain conditions this can lead to large deviations from the market equilibrium. We have previously shown that if M is allowed to affect the belief dynamics of agents, then the market spends most of the time in states corresponding to excess demand or excess supply. This kind of two-phase behavior [7] points to the destabilizing effect of apparently innocuous information exchanges in the market.

Very recently, we have observed that the collective behavior can also be destabilized if, instead of affecting the belief dynamics, the knowledge of M is used in evolving the structure of interactions J^t_{ij} between neighboring agents. This is implemented by assuming that agents seek out the most successful agents in its neighborhood, and choose to be influenced by them preferentially. Here, *success* is measured by the fraction of time the agent's decision (to buy or sell) accorded with the market behavior. As a rise in excess demand of an asset is taken to signal its desirability, an agent is considered successful if it is in possession of an asset that is in high demand. If an agent i is successful in predicting the market (i.e., its action in the last round accorded with the majority decision of the collective) then its interaction structure is unchanged. Otherwise, its neighboring agents with higher success are identified and the link strength between them, J_{ij}, is adjusted by an amount that is proportional to the ratio of the success of agent j to agent i. This implies that agents with higher success affect the decision process of agents with less success, but not the other way around. Finally, J_{ij} is normalized such that, for each agent, $\Sigma_i J_{ij} = 1$.

Fig. 1 (left) shows the resulting time series of the fractional excess demand. As the price P of the asset is governed by the demand for it, we can take P to be linearly related to M. This allows us to quantify the price fluctuations for the asset by calculating the logarithmic return of P as $R^t = \ln P^{t+1} - \ln P^t$. It is evident that the fluctuations are much larger than what would have been expected from an uncorrelated random process. This is further established when we plot the distribution of the return, normalized by its standard deviation, and compare it with the case where the J_{ij} are constant in time (Fig. 1, right). While the latter case is consistent with a Gaussian distribution, the model with adaptive interaction dynamics is found to exhibit a return distribution that has a power law tail. Moreover, the exponent of the cumulative distribution, $\alpha \simeq 3$, is found to agree *quantitatively* with the corresponding values observed in actual markets [8].

3 Conclusions

The observation of large price fluctuations (most strikingly during bubbles or crashes) implies that markets often display instabilities where the demand and supply are not even approximately balanced. We have seen in this paper that this is not necessarily inconsistent with the assumption that individual economic agents are rational. A simple agent-based model, where the structure of interactions between agents evolve over time based on information about the market, exhibits extremely large fluctuations around the market equilibrium that qualitatively match the fluctuation distribution seen in real markets.

References

1. Albin PS (1998) Barriers and bounds to rationality. Princeton University Press, Princeton
2. Chancellor E (1999) Devil take the hindmost: A history of financial speculation. Macmillan, London
3. Garber PM (1990) J. Economic Perspectives 4:35–54
4. MacKay C (1852) Memoirs of extraordinary popular delusions and the madness of crowds. National Illustrated Library, London
5. Iori G (2002) J. Economic Behavior & Organization 49:269–285
6. Sinha S, Raghavendra S (2004) SFI Working Paper 04-09-028
7. Sinha S, Raghavendra S (2005) in: Takayasu H (ed) Practical fruits of econophysics. Springer, Tokyo :200-204
8. Gopikrishnan P, Meyer M, Amaral LAN, Stanley HE (1998) Eur. Phys. J. B 3:139–140

Agent-Based Modelling with Wavelets and an Evolutionary Artificial Neural Network: Applications to CAC 40 Forecasting

Serge Hayward

Department of Finance Ecole Supérieure de Commerce de Dijon, 29, rue Sambin, 21000, Dijon, France
shayward@escdijon.com

Summary. Analysis of separate scales of a complex signal provides a valuable source of information, considering that different financial decisions occur at different scales. Wavelet transform decomposition of a complex time series into separate scales and their economic representation is a focus of this study. An evolutionary / artificial neural network (E/ANN) is used to learn the information at separate scales and combine it into meaningfully weighted structures. Potential applications of the proposed approach are in financial forecasting and trading strategies development based on individual preferences and trading styles.

1 Data generating mechanism

A recent move in financial research from homogenous to heterogeneous information added realism to the analysis of market efficiency. At the same time information heterogeneity does not necessarily imply and lead to beliefs heterogeneity. It is in fact the latter that seems to determine the empirical facts observed in financial markets. Unlike fully revealing equilibrium of homogeneous beliefs, in the environment with heterogeneous beliefs prices are driven by prevailing expectations of market participants. Thus forecasting future prices one must form expectations of others forecasts. Evolution of agents expectations to a large extent governs the adaptive nature of market prices. Overlapping beliefs of heterogeneous agents largely prevent the effective examination of expectation formation and price forecasting by traditional (time-series) methods.

The objective is thus to decompose a time series into a combination of underlying series, representing beliefs of major clusters of agents. Isolating effectively nonstationary and nonlinearly features, analysis of individual parts is expected to improve statistical inference. Emergent local type of behaviour is anticipated to be more receptive to forecasting. Weighted combination of individual forecasts is determined and evolves in accordance with specific market conditions, providing an aggregate forecast. Such multiscale analysis is based on multiresolution framework, focusing on the development of efficient numerical algorithms for signals decomposition and reconstruction.

2 Traders' heterogeneity

Economic agents behaviour is determined by their intrinsic opportunity sets. One of the main factors affecting traders opportunity sets is the time dimension, realised in differentiating market participants, according to their time horizons, into short and long term traders. Agents with distinct time scale dimensions respond in a different way, with different reaction time to the news. Low frequency shocks are likely to affect all market participants, though with some time delay. On the other hand high frequency shocks may be ignored, at least for a while, by long term traders. Such information processing inefficiency is heterogeneity-driven asymmetry by the market efficiency hypothesis. Identified casual relationships are potentially exploitable in trading model development.

Consider a scaling law that relates traders time horizons and frequency of price fluctuations. In these settings rapid price movements that corresponds to high frequency fluctuations are reflected in the frequent trading positions revaluations by agents with short term horizons, such as (inter-day) speculators. By trading upon the information of high frequency signals short term traders frequently execute transactions, supplying high frequency information to the market.

On the other hand slow price movements, corresponding normally to larger price shifts, are more apparent in low frequency signals, with less noise interference. Thus traders with long term horizons (e.g. certain institutional investors) tend to trade upon the information of low frequency signals. Long term traders, reducing their risk exposure with derivatives, stop-loss limits and the like act, to a certain extent, according to the market fundamentals [1]. Long term agents, trading upon low frequency signals, provide low frequency information to the market that is used adaptively by traders with similar time scale.

The interactions between heterogeneous agents in relationship to each other actions rather than to the market news produce endogenous dynamics in the market. Such dynamics provide reasonable explanation to some common empirical facts in Finance, as trend persistence or volatility clustering. Differentiating economic agents' expectations according to their time dimension has valuable consequences for forecasting. Since the time scale of traders is the key characteristic of the market, the adaptive dynamics of prices reflect beliefs and behaviour of the dominant agents on the market. Therefore, establishing a (trading) frequency signal of the dominant agents can be used for tuning into it and exploiting it in profitable predictions.

3 Wavelet transforms

Representing a signal in the *time-scale*[1] domain provides a fine time resolution, being narrow in time, to zoom in high frequencies with less ambiguity regarding their exact position in time. On the other hand a good frequency resolution is achieved by being short in frequency (long in time) with less ambiguity about the exact value of the frequency (to deal with low frequencies). To accomplish this, a signal is multiplied with the *wavelet* function[2] and the transform is computed for different segments of TD signal. Well localized in time, without constant resolution and infinite duration, wavelets adapt the time and frequency support, reaching the boundary of Heisenberg Uncertainty Principle[3].

By applying modified versions of a prototype, the *mother wavelet* (MW) o the time series, wavelet transform convolves the data with a series of local waveforms to discover correlated features or patterns. MW is a source function, from which translated and scaled wavelet functions (with different regions of support) are constricted. Nonorthogonal wavelet functions are used with both Continuous Wavelet Transform (CWT) and Discrete Wavelet Transform (DWT), whereas orthogonal[4] wavelet basses imply exclusively DWT.

4 Multiresolution analysis

The multiscale analysis utilises filters to split up a function into the components of subspaces and their orthogonal complements, representing different scales. Considering filtering as user-defined extraction of partial information from the input signal, a filter is an algorithm that divides a time series into individual components. In discrete setting a pair of sequences, $\{h(k); g(k)\}_{k \to Z}$ represents the lowpass and highpass filters. $h(k)$ smoothes the data, keeping low frequencies (longer-terms structures), while $g(k)$ preserves the detailed information, high frequencies (transitory patterns). The above filters are related to each other in the following way

$$g(n) = (-1)^n h(1-n). \tag{1}$$

Multiresolution analysis provides an intuitive understanding of the different levels of a time series, representing separate components of the underlying DGM. Decomposed series is a combination of an approximation (mean) and detail information at specified levels of resolution. Thus economic meanings can be attributed to different levels of approximation and details.

[1] Consider scale is an inverse of frequency.

[2] A wavelet refers to a complex function (with a zero integral over the real line) of a small wave, i.e. compactly supported (finite length) oscillatory function.

[3] Heisenberg Principle imposes limits on precising an event appearance with arbitrary accuracy in both time and period terms.

[4] Orthogonal wavelet functions characterised by no overlapping or projection.

5 Significance testing

Peak-based critical limit significance is used for signal analysis/processing. Two backgrounds are considered: a white noise, $H_0 : AR(1) = 0$ and a red noise (the signal power decreases with increased frequency), $H_0 : AR(1) > 0$. A 95% (99.9%) peak-based critical limit implies that in 1 out of 20 (1000) random noise signals would the largest peak reach this height by a random chance. Monte Carlo simulation is developed to generate the peak-based critical limits[5]. The simulated data is then fitted to bivariate, univariate or trivariate polynomials, depending on the number of factors affecting the significance.

6 Data analysis

The data considered includes prices for CAC 40 share index measuring the evolution of a sample of 40 equities listed on Euronext regulated markets in Paris. Created in June 1988, the CAC 40 is designed to represent the French stock exchange listed companies and serve as a support for derivatives markets. During the trading session 9.00 17.30, index levels are calculated in real time using the last trade quoted[6] and are disseminated every 30 seconds. Thus the change in the index is equal to the sum of the change in each of component equity times its weight in the index, where the market price of equity is adjusted for corporate actions taking effect, but not for dividend paid.

The period under investigation runs from 01.03.90 through 07.03.05 with the business time scale, which excludes holidays, Saturdays and Sundays from the physical time scale used in the experiment. The length of the data series is driven by the objective to explain the present behaviour of the index, where the data prior to 1990 is considered to refer to a different from the current phenomena. The original data, consisting of the series with tick[7] frequency of 30 seconds was obtained from Euronext.

Considering that financial markets are populated by heterogeneous agents, the dominant traders behaviour determines the adaptive nature of the prices on that market. Under such approach, one needs to identify the most appropriate sampling intervals for the analysis and predictions of the dynamics in the market under consideration. A selected sampling interval should reflect the information about the (dominant) traders reaction time to the news and

[5] The Monte Carlo method is used to determine significance levels for a process with unknown statistical distribution.

[6] On the exceptional basis in the event that at least 35% of the market capitalization of the index has not been traded, a price indicator - the forerunner is substituted for the index.

[7] A 'tick' is one logical unit of information.

their dealing frequency. At the same time for this information to be apparent for the analysis the amount of (random) noise in the series ought to be minimise.

Through determining the most prominent (informative) frequency bands in the original signal the appropriate sampling interval is identified. Examining the DWT coefficients using Daubechies wavelet basses with the maximum levels of decomposition, the main energy is found to be around the scale 2^10 in dyadic scaling, which corresponds to the frequency between 8 and 9 hours. This result justifies the data subsampling with 8.5 hours frequency that produce a relatively low noise signal without significant loss of information.

Thus from the original data after subsampling, the series containing 3779 eight and a half hours prices was extracted. The analysis of such frequency well relates to the objective of examining the behaviour of heterogeneous agents that strategically fulfil certain goals and cluster according to some (nontrivial) time horizons, optimal for their economic type. [2] assert a similar sampling frequency as the most appropriate in designing an ANN by a typical off-floor trader. Furthermore, for practical purposes of construing a model, generating reliable predictions, the issues of the realistic time needed to execute a strategy makes the choice of such a sampling interval justified.

7 Low and high frequencies

Assuming that lower level decomposition capture the long range dependencies whereas the higher levels identify short term relationships, the signal is decomposed into low and high frequencies. Experiments with different decomposition techniques, namely CWT, DWT and WPT demonstrate that different transforms are appropriate for the data with different sampling frequencies. Specifically it was found that for a high frequency signal the best results were achieved by DWT with Daubechies wavelet basses, whereas for the signal with longer sampling interval the most promising results were with CWT using Gabor wavelet function. The results for high frequency signals confirm that moving averages and moving differences (given by DWT coefficients) are useful tools for identifying short term dependencies in the noisy environment. In addition it appeared to be an effective data reduction scheme working with large data sets, due to DWT effective subsampling operations. The efficiency of the pyramid algorithm of [3] in the Daubechies-Mallat paradigm is ideal for rapid processing in real time. Smooth and continuous longer range relationships are more susceptive to CWT that provides a greater accuracy. Disjunct nature of Daubechies wavelets contrast with smooth motions that underline the data. Furthermore, being computationally intensive CWT appears to be appropriate for small and medium size data sets. Inferior results for Wavelet Packet Transform were found with high and low frequency signal considered. A further research into the behaviour of Wavelet Packet Transform coefficients is required before any conclusion can be drawn on this matter.

To capture and exploit the features of interest with minimum delay to be of value for financial applications, CWT with Gabor wavelet function generated the best performance due to its ability of efficient filtering with minimal lag. A number of techniques to reduce lag are considered. The *Slow taper* approach proposes to increase the frequency domain period and reduces the frequency domain width. Under a worse case scenario the lag is equal to the width (reduced by the slow taper). Another approach considered is to *truncate the filter* before its effective response has dropped to zero. *Mirroring* as a lag reduction technique appends historical points in reverse order. Without knowing the TD impulse response, it simulates a half-impulse filter with all weights doubled, except the highest (the most recent) ones. Being straightforward to implement mirroring comes at cost for asymmetric signals[8].

Considering a lowpass filter data smoothing, the best performance was obtain by the Gabor wavelet ($\mu = 350$; $p = 350$; $l = 350$) for both price and return series. Balancing spurious frequency responses against lag reduction benefits (real) mirroring technique was found to be the most effective for the current application.

In high frequency filtering a highpass real Gabor ($\mu = 3$; $p = 5$; $l = 2$) generates the most accurate result for both price and return series. Applying a very narrow in TD filter, Morlet shape modification was used to avoid contamination from slow trends. Such modification centres the filter and eliminate/reduce the problem.

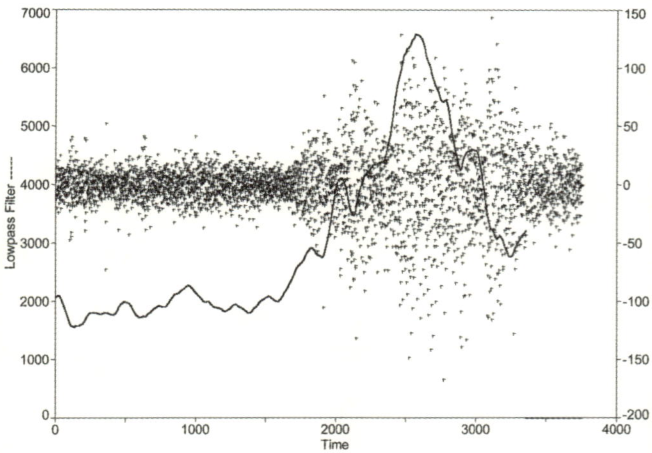

Fig. 1. Low and High Frequencies. Low frequency - solid curve; - high frequency - dotted curve.

[8] To overcome problems related to the extension of the time-domain wavelet functions over the entire temporal domain in discrete wavelet analysis settings Pollock (2004, 2005) propose to use wrapping the wavelet or filter coefficients around a circle of circumference and adding the overlying coefficients.

The memory analysis of two series obtained indicates that low frequency signal now curries most of the memory (the Hurst exponent, H = 0.098). Low persistency for high frequency signal is confirmed by the value of H close to 0.5. These results indicate that the original data series is successfully decomposed into signals currying separately main characteristics and allowing their individual analysis and modelling without undesirable interferences from each other.

Considering the individual outputs of the lowpass and highpass filters, the experiment detect that low and high frequencies move into and out of phase with each other. Figure 1 presents low and high frequencies on the same plot[9]. Around 03-09.00 [2501-2648] and 03-04.03 [3262-3302] one can see apparent phase shifts, rather then structural breaks, as it was claimed previously [4].

8 Multiscale prediction

The prediction scheme used in this study, steaming from the autoregressive (AR) methodology, is based on extracting important occurrences (such as turning points) from the data to determine the time of their possible reappearance in the future. In AR model fitted to individual scales of multiresolution transform, wavelet coefficients are used as a series of independent variables to forecast future values of the variable of interest. If the DGP is of autoregressive nature, the prediction converges to the optimal forecast and is asymptotically equivalent to the best forecast. ANN then learns the behaviour of the time series by changing the weights of the interconnected neurons comprising the network.

A heterogeneous beliefs model with expectations differentiated according to the time dimension is developed through decomposing a time series into a combination of the underlying series, representing beliefs of major clusters of market participants. In adaptive analysis of local behaviour of heterogeneous agents the high and low frequencies signals are distinguished to represent the short and long term traders. By separately investigating different frequency frames, the aim is to identify the dominant cluster of traders on the market considered and the adaptive nature of such market prices. Using observed values for each frequency, equations are evolved, resulting in separate trading signals for low (S-L), high (S-H) and combined (S-LH) frequencies.

In Table 1, $P_{t-i}^{LF}, P_{t-i}^{HF}$ are price, $fc(P_t^{LF}), fc(P_t^{HF})$ are forecast and R_t^{LF}, R_t^{HF} are return series for high and low frequencies respectively. Weighted combination of individual frequencies provides the aggregate signal, S-LH. Weighting of individual frequencies is determined by GA and is specific for the market under investigation. As benchmarks two strategies are considered: buy and hold strategy (B/H) and a strategy based on undecomposed price series (S-P) with independent variables identical to those presented in Table 1.

[9] Note that the graphs are shifted so as to cause the features to align in a visually obvious manner.

Table 1. Trading signals

S-L $= \{P^{LF}_{t-i}, fc(P^{LF}_t), R^{LF}_t\}$,
S-H $= \{P^{HF}_{t-i}, fc(P^{HF}_t), R^{HF}_t\}$,
S-LH $= \{P^{LF}_{t-i}, P^{HF}_{t-i}, fc(P^{LF}_t), fc(P^{HF}_t), R^{LF}_t, R^{HF}_t\}$.

For this experiment the structure developed in Hayward (2005) is adopted. A posterior optimal rule signal[10] is modelled with an EANN. A dual ANN consists of forecasting network feeding into the acting network. The information set of the acting network includes the output (one period ahead) of the forecasting (AR) network, the inputs used for forecasting (to relate the forecast to the data upon which it was based), as well as the log return, (given by $r(t) = r(\Delta t; t) = x(t)x(t\Delta t)$, where $x(t)$ is a price series). In evolutionary settings a population of networks, representing agents facing identical problems but generating different solutions, is assessed on the basis of their fitness. A single hidden layer Time-Lag Recurrent Network with the number of the hidden layers neurons determined genetically and GA with tournament selection, size 4; probability of uniform mutation [0, 0.05] and probability of uniform crossover [0.7, 0.95] are used in the experiment.

To simulate 'true' forecasting and actual trading, the performance of the models evaluated on previously unseen data. To compare the developed models among themselves, the identical out-of-sample period is used and annualised results are presented. To achieve compatibility with similar studies, the internal error, determined by MSE loss function is produced, comparing ANN output to the desired response, given by the next period price in forecasting and the current strategy in signal modelling. Directional accuracy (DA) is used for performance surface optimisation with GA. Optimising the learning the sign of the desired output is adopted for the reason of its established links with profitability [6].

The experiment considered 75 forecasting and 120 trading strategies settings with multiple trials run for each settings. GA was capable to identify 'optimal' settings on the average in 80% of 10 individual runs. By simulating the traders price forecast and their trading strategy evolution, eleven strategies were able to outperform the B/H benchmark in economic terms with an investment of Euro 10,000 and TC of 0.3 of trade value. Those eleven strategies are represented by five strategies based on combined low and high frequencies; three on high frequency signal; two on low frequency and one strategy developed with undecomposed price series. Average return improvement over the B/H benchmark is 550% for S- LH; 370% for S-H; 74% for S-L and 20% for S-P.

[10] The rule, using future information to determine the best current trading action, returns a buy/sell signal if prices tomorrow have increased/decreased.

Table 2. Statistical Characteristics: [Accuracy: percentage of correct predictions; Correlation: between desired and ANN output].

Stats./Signals	S-P	S-L	S-FH	S-ALH
Accuracy (%)	58.3	50.7	43	34
Correlation	0.51	0.58	0.02	0.21
MSE Error	0.85	0.71	1.01	1.03

Statistical characteristics, presented in Table 2 identify the best performance in forecasting and signal modelling by the unfiltered price series and low frequency signal. At the same time the statistical measures for the two most profitable strategies are far from satisfactory. These results confirm that economic profitability has only a weak relationship with common statistical measures. Therefore selecting a trading model on the basis of its statistical characteristics would result in overlooking potentially profitable models. Simulation with EANN provides a reliable platform for off-line models testing.

Table 3 presents out-of-sample economic performance of the main trading signals. The primary and secondary (in profitability terms) signals outperform S-P benchmark by 542% and 392% respectively. The risk exposure, measured by the Sharpe ratio, demonstrates that the primary and secondary strategies are respectively 15.75 and 14.5 times less risky than S-P benchmark. The largest percentage loss occurred during open trades, given by the Maximum Drawdown indicates that two primary strategies characterised by considerably lower downside risk. Superior performance of S-L and S-P signals in the percentage of winning trades should be viewed together with their infrequent trading and high downside risk exposures. The primary and particular secondary signals are characterised by high number of trades, indicating that those signals are exploited by short term speculators.

Even accounting for transaction costs of 0.3% , the extra returns achieved with the primary and secondary signals make those strategies the most profitable despite their high trading frequencies. The dealing frequency of the most active traders (making positive profit) confirms the right choice of the sampling frequency adopted in the experiment and so no re-sampling is necessary. For practical applications the information about dealing frequencies of profitable simulated traders could be used to readjust sampling interval (previously selected by some rule or even ad hoc).

Consider the length of trades, measured by the average number of bars the given type of trade is active. S-L signal is characterised by the longest holding periods, confirming that this signal is used by long term investors. On the other hand S-H signal display very short length of trades that might be particular suitable for short term speculators. Notice that a cumulative signal S-LH is also characterised by short holding periods suggesting that the market is likely to be dominated by short term traders. Such information about the behaviour of the dominant traders on the market under investigation is of a great value for predicting adaptive nature of the market prices. Furthermore, providing good guidance on how actively a signal needs to be traded such experiment can also be considered for the user making a decision to tune into the cycle best suited to an individual trading style.

Table 3. Economic Performance

	S-LH (Primary)	S-H (Second.)	S-L	S-P
Return Annual.	35.10	25.38	9.40	6.48
Sharpe Ratio	0.63	0.58	0.2	0.04
Profit/Loss	10,87	6,81	3.3	2.03
Max. Drawdown	-0.2	-1.94	-23.27	-14.19
Annual Trades	81	122	5	11
% Win. Trades	75.7	61.7	88.6	76.8
%Winning up Periods	49.77	50.28	50.19	47.53
Length of Trades Win/Loss/Out of Market	3/2/1	1/2/1	147/44/3	29/21/109

Annualised Return: $r = 1/N \sum_{t=1}^{N} R_t \times 252$;
Sharper Ratio: $\mathrm{SR} = r/\sigma$;
Max. Drawdown: $\min_{i=1,\dots,t=1,\dots,N} \left(\sum_{j=1}^{N} X_j \right)$;
% Winning Trades: WT = Number of $R_t > 0/$ (Total Number of Traders)$\times 100$

Finally observing the length of each signal out of market, one can see that all signals based on the decomposed series spend very little time in out of market position in comparison to the signal based on undecomposed price series. This result is attributed to the improved abilities of EANN to learn information from signals decomposed on different frequencies. Comparing the signal constructed by a weighted combination of high and low frequencies with the signal developed using the price series, the former is more informative spending very time out of market (being unable to predict market behaviour).

EAAN display good generalisation abilities, as indicated by percentage of winning up periods close to a half for most models considered. EAAN has successfully learned the in-sample information (constructed to be representative of upward and downward movements). It also displays an adequate performance in predicting a fall and a rise on out-of-sample data. To examine the stability of the proposed model different in/out-of-sample periods were considered. In particular the memory of the lengths of visually acknowledged complete cycles were adopted in the experiment, signifying that previous cycle behaviour might not be necessary informative for modelling of the current state. Only slight improvement in profitability was observed, given by average increase in annualised return by 1- 3% and in Sharpe ratio by 2-3%. Thus it appears that decomposing the time series on the frequency signals (representing the time dimension of the major clusters of the market participants) is more effective for profitable models development than the search for the 'optimal' in and out of sample length (representing traders past and forward time horizons). In particular wavelet transforms, lending themselves to shifting along time in a consistent way, reveal features that have appeared in the past (without the necessity to determine with reliable precision the appropriate length of this past).

GA optimization results in average improvement in annualised return by 70%; Sharpe ratio riskness by 35%; and P/L by 270%. Optimized signals trade consistently less frequently with higher percentage of winning trades. At the same time the downside risk, measure by ξ is on average twice higher for signals optimised with GA. The last observation illustrates certain limitations with applications of computational intelligence in finance. Results obtained sometimes lacking clear reasoning. This shortcoming is often exacerbated by restrictions on using common statistical inference techniques (e.g. significance testing) with resultant dependence on heuristics.

Examining the weighting of the aggregate strategy S-LH the analysis identifies that high frequency signal dominate significantly the low frequency. Experiments with weights modification results in annualised return decrease, indicating that EAAN was capable to find the optimal frequencies combination. Therefore the market considered appears to be dominated by agents, trading primarily on the high frequency signals. The contribution of the agents using the low frequency although small, is not negligible. The emergent from signals decomposition local type of behaviour is proven to be more receptive to forecasting and modelling.

9 Conclusion

Assuming that lower level decomposition capture the long range dependencies whereas the higher levels identify short term relationships, the signal is decomposed into low and high frequencies. Considering the individual outputs of the lowpass and highpass filters, the experiment detect that low and high frequencies move into and out of phase with each other. Two apparent phase shifts (rather then structural breaks, as it was claimed previously) were identified.

A heterogeneous beliefs model with expectations differentiated according to the time dimension is developed through decomposing a time series into a combination of the underlying series, representing beliefs of major clusters of market participants. In adaptive analysis of local behaviour of heterogeneous agents the high and low frequencies signals are distinguished to represent the short and long term traders. The results of the experiment demonstrate improved abilities of EANN to learn information from signals decomposed on different frequencies. Comparing the signal constructed by a weighted combination of high and low frequencies with the signal developed using the price series, the former is more informative spending very time out of market (being unable to predict market behaviour). Examining the primary strategy, the analysis identifies that high frequency signals dominate significantly the low frequencies. Therefore the market considered appears to be dominated by agents, trading primarily on the high frequency signals. Though the contribution of the agents using the low frequency being small is not negligible. The emergent from signals decomposition local type of behaviour is proven to be more receptive to forecasting and modelling.

To examine stability of the proposed model different in/out-of-sample periods were considered. In particular the memory of the lengths of visually acknowledged complete cycles were adopted in the experiment, signifying that previous cycle behaviour might not be necessary informative for modelling of the current state. Only slight improvement in profitability was observed. Thus it appears that decomposing the time series on the frequency signals (representing the time dimension of the major clusters of the market participants) is more effective for profitable models development than the search for the optimal in and out of sample length (representing traders past and forward time horizons). In particular wavelet transforms reveal features that have appeared in the past (without the necessity to determine with reliable precision the appropriate length of this past).

References

1. Dacorogna M. et al.(2001) An Introduction to High-Frequency Finance. Academic Press
2. Kaastra I, Boyd M (1996) Neurocomputing 10:215–236
3. Mallat SG (1989) IEEE Transactions on Pattern Analysis and Machine Intelligence 11(7):674–693
4. Hayward S (2005) Computational Economics 25(1-2):25–40
5. Leitch G, Tanner E (2001) American Economic Review 81:580–590
6. Hayward S (2006) in Practical Fruits of Econophysics, Ed. Takayasu H, Springer-Verlag: Tokyo. p. 99–106.

Information Extraction in Scheduling Problems with Non-Identical Machines

Manipushpak Mitra *

Economic Research Unit, Indian Statistical Institute, 203 B. T. Road, Kolkata-700108, India. mmitra@isical.ac.in

Summary. In this paper we study the problem of scheduling a set of jobs through a set of machines where the processing speed of the machines can differ. We assume that the waiting cost of each job is private information and that all jobs take identical processing time in any given machine. By allowing for monetary transfer, we identify the complete class of multiple non-identical machine scheduling problems for which we can find information revelation mechanisms that lead to (a) minimization of aggregate cost (or efficiency) and (b) costless information extraction.

1 Introduction

The literature on mechanism design under incomplete information is large and growing. In this context, a very robust notion is that of information extraction in dominant strategies where it is in the interest of each informed agent to reveal her private information irrespective of the announcement of other agent(s). The classic result is due to Vickrey [18], Clarke [2] and Groves [6]. They have designed a class of mechanisms, known as VCG mechanism that implements the efficient decision in dominant strategies when monetary compensation is allowed. In general, if preferences are quasi-linear, VCG mechanisms are sufficient to implement efficient decision in dominant strategies. Given quasi-linearity and "smoothly connected" domains, VCG mechanisms uniquely implement the efficient decision in dominant strategies (see Holmström [9]). The main drawback of VCG mechanisms is budget imbalance (see Green and Laffont [5], Hurwicz and Walker [10] and Walker [19]). The damaging nature of budget imbalance, in the public goods context, was pointed out by Groves and Ledyard [7]. They proposed, using a very simple model, that an alternative procedure based on majority rule voting may lead to an allocation of resources which is Pareto superior to the one produced by the VCG mechanism.

* The author is grateful to Hervé Moulin for suggesting this problem.

In the public goods problem, Groves and Loeb [8] have proved that if preferences are quadratic then we can find balanced VCG mechanisms. This result was generalized by Tian [17] and Liu and Tian [12]. Moreover, for *single machine* scheduling problems with linear costs, we can find balanced VCG mechanisms (see Mitra [13], [14] and Suijs [16]). In the *multiple identical machine* scheduling problem with linear costs, Mitra [15] provides a sufficient class of problems for which we have the same possibility result. In this paper we ask the following question: Can we extend this possibility result to *multiple non-identical machine* scheduling problems? By multiple non-identical machine scheduling problem we mean the following setup: (a) there are n agents and m machines, (b) each agent has exactly one job to process using these machines, (c) each machine can process one job at a time, (d) jobs are identical in the sense that all jobs have same completion time in any given machine, and (e) machines are non-identical in the sense that completion time for any given job may differ across machines.

The multiple non-identical machines scheduling problem, defined above, resembles some of the scheduling problems that are analyzed in the operations research literature. Papers relating to scheduling n jobs through m machines by Dudek and Teuton Jr. [4], flow shop scheduling problems with ordered processing time by Dudek, Panwalkar and Smith [3] and flow shop problems with dominant machines by Krabbenborg, Potter and van den Nouweland [11] deal with finding algorithms to order (or queue) the n jobs through m machines in an efficient way (that is, in a way that minimizes the total elapsed time). Unlike these papers, our primary aim is to address the issue of information revelation in order to implement the efficient order.

2 The model

Let $N = \{1, \ldots, n\}, n \geq 2$ be the set of agents and $M = \{1, \ldots, m\}$ be the set of machines. Each agent has one job to process. Agents can be served only sequentially in different machines. Serving any agent takes same amount of time in any given machine. Given any machine $q \in \{1, \ldots, m\}$, the speed of completing a job is $s_q \in (0, 1]$. Therefore, the processing speed of different machines is captured by the vector $s = (s_1, \ldots, s_m) \in (0, 1]^m$. We assume without loss of generality that $0 < s_1 \leq \ldots \leq s_m = 1$. Each agent is identified with a waiting cost $\theta_i \in \Re_+$, her cost of waiting per unit of time. A typical profile of waiting cost is $\theta = (\theta_1, \ldots, \theta_n) \in \Re^n_+$.

The server's objective is that of efficiency. Efficiency implies that, given any profile $\theta \in \Re^n_+$, the server schedules n jobs through m machines in order to minimize the aggregate waiting cost. To achieve this goal, the server has to consider two things. First, the server has to select the vector of smallest n numbers from the set $\{ks_1, \ldots, ks_m\}_{k=1}^n$ and second, given this selection, the server has to arrange the jobs in a non-increasing order of their waiting costs.

Let $z(s) = (z(1), \ldots, z(n))$ represent the smallest n elements obtained from the set $\{ks_1, \ldots, ks_m\}_{k=1}^n$. We assume without loss of generality that $0 < z(1)(\equiv s_1) \leq \ldots \leq z(n)$. We refer to $z(s)$ as the completion time vector. Given any $s = (s_1, \ldots, s_m) \in (0,1]^m$, efficiency requires that the choice of $z(s)$ is unique for all realizations of $\theta \in \Re_+^n$. An algorithm to find the $z(s)$ vector given $s = (s_1, \ldots, s_m)$ can be found in Brucker [1]. The server then ranks the n jobs and assigns them to $z(s) = (z(1), \ldots, z(n))$. A rank vector is an onto function $\sigma : N \to N$. If agent i's rank is σ_i under some rank vector $\sigma = (\sigma_1, \ldots, \sigma_n)$, then her total cost is $z(\sigma_i)\theta_i$. Let $\Sigma(N)$ be the set of all possible rank vectors of agents in N.

Definition 1 A rank vector $\sigma \in \Sigma(N)$ is *efficient* for a profile $\theta \in \Re_+^n$ if $\sigma \in \operatorname{argmin}_{\sigma \in \Sigma(N)} \sum_{i=1}^n z(\sigma_i)\theta_i$.

Given $z(s)$, the implication of efficiency is that agents are ranked according to the non-increasing order of their waiting costs (that is, if $\theta_i \geq \theta_j$ under a profile θ, then $z(\sigma_i) \leq z(\sigma_j)$). Moreover, there are profiles for which more than one rank vector is efficient. For example, if all agents have the same waiting cost, then all rank vectors are efficient. Therefore, we have an efficiency correspondence. In this paper we consider a particular efficient rule (that is, a single valued selection from the efficiency correspondence). For our efficient rule, we use the following *tie breaking rule*: if $i < j$ and $\theta_i = \theta_j$ then $\sigma_i < \sigma_j$. This tie breaking rule guarantees that, given a profile $\theta \in \Re_+^n$, the efficient rule selects a single rank vector from $\Sigma(N)$.

If the agents' waiting costs are common knowledge then the server can easily implement the efficient rule. However, if the waiting cost is private information, then agents have an incentive to misreport. In particular, in the absence of any monetary transfers, the agents have an incentive to overstate their waiting cost so that they are served quickly. The server's problem then is to design a mechanism in order to extract information. We refer to the multiple non-identical machine scheduling problem as $G = (N, M, s, \Re_+)$. Let $\mathbf{G}$ denote the set of all possible multiple non-identical machine scheduling problems such that $|N| = n < +\infty$ and $|M| = m < +\infty$ and $s \in (0,1]^m$ with $0 < s_1 \leq \ldots \leq s_m$.[2] Given a $G = (N, M, s, \Re_+) \in \mathbf{G}$, let $\theta = (\theta_1, \ldots, \theta_n) \in \Re_+^n$ be an announced profile of waiting costs. A mechanism μ associates to each announced profile θ, a tuple $(\sigma(\theta), t(\theta)) \in N \times \Re^n$ where $\sigma_i(\theta)$ is the rank of agent i and $t_i(\theta)$ is the corresponding transfer. Let $u_i(\sigma_i(\theta), t_i(\theta), \theta'_i) = -z(\sigma_i(\theta))\theta'_i + t_i(\theta)$ be the utility of agent i when the profile θ is reported (collectively) and her true waiting cost is θ'_i. Given a profile θ, let θ_{-i} denote the profile $(\theta_1, \ldots, \theta_{i-1}, \theta_{i+1}, \ldots, \theta_n)$.

Definition 2 A mechanism (σ, t), for $G \in \mathbf{G}$, is *dominant strategy incentive compatible* (or DSIC) if $\forall\, i \in N$, $\forall\, \theta_i, \theta'_i$ and $\forall\, \theta_{-i}$,

$$u_i(\sigma_i(\theta), t_i(\theta); \theta_i) \geq u_i(\sigma_i(\theta'_i, \theta_{-i}), t_i(\theta'_i, \theta_{-i}); \theta_i).$$

[2] Here $|X|$ denotes the cardinality of the set X.

DSIC guarantees that truth telling is a dominant strategy for each $i \in N$ irrespective of the announcement of $N\backslash\{i\}$.

Definition 3 A mechanism (σ, t), for $G \in \mathbf{G}$, is *efficient* (or EFF) if for all announced profiles θ, $\sigma(\theta) \in \text{argmin}_{\sigma \in \Sigma(N)} \sum_{i=1}^{n} z(\sigma_i)\theta_i$.

A mechanism is EFF if it selects the efficient rank vector contingent on the announced profile $\theta \in \Re_+^n$.[3] If, for $G \in \mathbf{G}$, a mechanism achieves EFF and DSIC then we say that, for G, efficient decision is implementable in dominant strategies.

Definition 4 A mechanism (σ, t), for $G \in \mathbf{G}$, is *budget balanced* (or BB) if for all announced profiles θ, $\sum_{i=1}^{n} t_i(\theta) = 0$.

If, for $G \in \mathbf{G}$, a mechanism satisfies EFF, DSIC and BB then we say that the problem $G \in \mathbf{G}$ is ex-post (or first best) implementable. With first best implementability, information extraction is costless.

3 Results

The following result is a special case of Theorem 1 of Holmström [9].

Proposition 1 A mechanism (σ, t), for $G \in \mathbf{G}$, is EFF and DSIC *if and only if* for all $\theta \in \Re_+^n$,

$$t_i(\theta) = -\sum_{j \neq i} z(\sigma_j(\theta))\theta_j + g_i(\theta_{-i}) \text{ for all } i \in N. \quad (1)$$

Proof: Follows from Theorem 1 of Holmström [9] since the domain of profiles $\Re_+^n$ is convex implying smooth connectedness. □

Remark 1 The mechanism with the transfer satisfying condition (1) is known as the VCG mechanism. The function $g_i(\theta_{-i})$ cannot depend on either the value θ_i or the rank $\sigma_i(\theta)$ of agent i. However, it can depend on θ_{-i} as well as on the ranks of agents in $N\backslash\{i\}$. Here $N\backslash\{i\}$ is the economy where agent i is not present.

The following result gives a complete characterization of the class of multiple non-identical machine scheduling problems for which one can find balanced VCG mechanisms. Consider the vector $\alpha_n = (\alpha(1), \ldots, \alpha(n))$ where $\alpha(k) = (-1)^{k-1}\binom{n-1}{k-1}$ for all $k \in \{1, \ldots, n\}$. Our result shows that for a $G = (N, M, s, \Re_+) \in \mathbf{G}$ one can find mechanism that satisfies EFF, DSIC and BB if and only if $z(s)$ is orthogonal to α_n.

[3] We use the same tie breaking rule to select a unique state contingent rank vector from the efficiency correspondence.

Theorem 1 A mechanism (σ, t), for $G = (N, M, s, \Re_+) \in \mathbf{G}$, is EFF, DSIC and BB *if and only if* $\alpha_n.z(s) = 0$.

Proof: We first prove the necessity, that is, if, for $G \in \mathbf{G}$, $\exists\ \mu = (\sigma, t)$ that satisfies EFF, DSIC and BB, then $\alpha_n.z(s) = 0$. Observe that if a mechanism is EFF and DSIC then from Proposition 1 we know that for all $i \in N$ and for all $\theta \in \Re^n_+$, $t_i(\theta) = -\sum_{j \neq i} z(\sigma_j(\theta)\theta_j + g_i(\theta_{-i})$. Furthermore, since transfers are balanced, we can use the result of Walker [19] to conclude that for all pairs of profiles $\theta, \theta' \in \Re^n$, we must have

$$\sum_{S \subseteq N} (-1)^{|S|} \sum_{i \in N} z(\sigma_i(\theta(S)))\theta_i(S) = 0 \tag{2}$$

where $\theta(S) = (\theta_1(S), \ldots, \theta_n(S)) \in \Re^n_+$ is a profile such that $\theta_j(S) = \theta_j$ if $j \notin S$ and $\theta_j(S) = \theta'_j$ if $j \in S$. Pick three distinct positive numbers a, b, c such that $0 < a < b < c$. Consider two profiles $\theta = (\theta_1, \ldots, \theta_n)$ and $\theta' = (\theta'_1, \ldots, \theta'_n)$ such that $\theta_1 = b$, $\theta_j = a$ for all $j \in N - \{1\}$ and $\theta'_i = c$ for all $i \in N$. Applying equation (2) to the pair θ, θ' and then simplifying it, we get $(b-c)\alpha_n.z(s) = 0$. Since $b \neq c$, the result follows.

To prove sufficiency, we show that if $G = (N, M, s, \Re_+) \in \mathbf{G}$ is such that $\alpha_n.z(s) = 0$, then we can find a balanced VCG mechanism. For that we have to find appropriate $g_i(.)$ functions. This is done in the following way. We first define a vector $H(s)$ of $(n-1)$ elements that satisfies a recursion relation with the elements of $z(s)$. Secondly, we define the rank vector of the set of $N\backslash\{i\}$ agents. Finally, using the $H(s)$ vector and the rank vector of $N\backslash\{i\}$ agents, we define the $g_i(.)$ function that gives a balanced VCG mechanism.

Consider the vector $H(s) = (h(1), \ldots, h(n-1))$ such that for all $k \in \{1, \ldots, n-1\}$,

$$h(k) = \sum_{r=1}^{k} \frac{(-1)^{k-r}\binom{n-1}{r-1} z(r)}{(n-1)\binom{n-2}{k-1}} \tag{3}$$

Claim 1: Given $\alpha_n.z(s) = 0$, the following recursion relation holds for all $k \in \{1, \ldots, n\}$.

$$z(k) = (n-k)h(k) + (k-1)h(k-1) \tag{4}$$

Proof: See Proposition 2.1 of Mitra [13].

For an arbitrary agent $i \in N$ and $\theta_{-i} \in \Re^{n-1}_+$ we define a rank vector, $\bar{\sigma}(\theta_{-i}) : N\backslash\{i\} \to \{1, \ldots, n-1\}$ as follows: for all $j, k \in N - \{i\}$, $\bar{\sigma}_j(\theta_{-i}) < \bar{\sigma}_k(\theta_{-i})$ if and only if either $\{\theta_j > \theta_k\}$ or $\{\theta_j = \theta_k \text{and} j < k\}$ hold. It follows from the construction of $\sigma(\theta)$ and $\bar{\sigma}(\theta_{-i})$ that the following property holds: for all $j \neq i$,

$$\bar{\sigma}_j(\theta_{-i}) = \begin{cases} \sigma_j(\theta) & \text{if } \sigma_j(\theta) < \sigma_i(\theta) \\ \sigma_j(\theta) - 1 & \text{if } \sigma_j(\theta) > \sigma_i(\theta) \end{cases}$$

The last condition follows from efficiency of the rank vector $\bar{\sigma}(\theta_{-i})$ for an economy with $N\backslash\{i\}$ agents. It simply guarantees that the following: in any state $\theta \in \Re^n_+$, if an agent $i \in N$ with waiting cost θ_i is removed, the relative ranking of waiting costs of $N\backslash\{i\}$ agents remain unchanged.

For any distinct $i, j \in N$ and $\theta_{-i} \in \Re^{n-1}_+$, $h_j(\bar{\sigma}_j(\theta_{-i}))$ is defined as in (3) that is, $h_j(\bar{\sigma}_j(\theta_{-i})) = \sum_{r=1}^{\bar{\sigma}_j(\theta_{-i})} \frac{(-1)^{\bar{\sigma}_j(\theta_{-i})-r}\binom{n-1}{r-1}z(r)}{(n-1)\binom{n-2}{\bar{\sigma}_j(\theta_{-i})-1}}$. We consider the sum $\sum_{i:i\neq j} h_j(\bar{\sigma}_j(\theta_{-i}))$ for agent j, that is, we fix j and sum over all $i \in N\backslash\{j\}$. Using the relationship between $\bar{\sigma}_j(\theta_{-i})$ and $\sigma_j(\theta)$ stated above and noting that $|\{i|\sigma_j(\theta) < \sigma_i(\theta)\}| = n - \sigma_j(\theta)$ and $|\{i|\sigma_j(\theta) > \sigma_i(\theta)\}| = \sigma_j(\theta) - 1$, we obtain

$$\sum_{i:i\neq j} h_j(\bar{\sigma}_j(\theta_{-i})) = (n-\sigma_j(\theta))h_j(\sigma_j(\theta)) + (\sigma_j(\theta)-1)h_j(\sigma_j(\theta)-1) = z(\sigma_j(\theta)). \tag{5}$$

The last equality in (5) follows from condition (4) in Claim 1.

For all $i \in N$ and $\theta_{-i} \in \Re^{n-1}_+$, define $g^*_i(\theta_{-i}) = (n-1)\sum_{j:j\neq i} h_j(\bar{\sigma}_j(\theta_{-i}))\theta_j$ Observe that given $g^*_i(.)$, $\sum_{i\in N} g^*_i(\theta_{-i}) = (n-1)\sum_{i\in N}\sum_{j\neq i} h_j(\bar{\sigma}_j(\theta_{-i}))\theta_j = (n-1)\sum_{j\in N}\sum_{i\neq j} h_j(\bar{\sigma}_j(\theta_{-i}))\theta_j = (n-1)\sum_{j\in N} z(\sigma_j(\theta))\theta_j$. Therefore,

$$\sum_{i\in N} g^*_i(\theta_{-i}) = (n-1)\sum_{j\in N} z(\sigma_j(\theta))\theta_j \tag{6}$$

Finally, consider the VCG mechanism (σ, t^*) where for all announced profiles $\theta \in \Re^n_+$, $t^*_i(\theta) = -\sum_{j\neq i} z(\sigma_j(\theta))\theta_j + g^*_i(\theta_{-i})$. Using (6) it follows that for (σ, t^*), $\sum_{i\in N} t^*_i(\theta) = -(n-1)\sum_{j\in N} z(\sigma_j(\theta))\theta_j + \sum_{i\in N} g^*_i(\theta_{-i}) = 0$. Thus, we have obtained a balanced VCG mechanism. □

Remark 2 The single machine problem with linear cost (Mitra [13]) is a special case of the multiple non-identical machine scheduling problem. In particular, with single machine, $M = \{1\}$, $s = (s_1 = 1)$ and $z(s) = (z(1), \ldots, z(n))$ is such that $z(k) = k$ for all $k \in \{1, \ldots, n\}$. It is easy to verify that, as long as $n \geq 3$, the completion time vector $z(s)$ is orthogonal to α_n. For $n = 2$, $\alpha_2.z(s) = -1 \neq 0$ and we have an impossibility result.

Remark 3 If $n = 2$ and $m \geq 2$, then $\alpha_2.z(s) = 0 \Rightarrow z(1) = z(2) \Rightarrow s_1 = s_2$. Then the problem is trivial as two agents have no incentive to misreport.

Remark 4 The multiple identical machine problem with linear cost (Mitra [15]) is also special case. For the multiple identical machine problem, $M = \{1, \ldots, m\}$, $m \geq 2$, $s_1 = \ldots = s_m = 1$ and for all $k \in \{1, \ldots, n\}$, $z(k) = \left[\frac{k}{m}\right]_+$.[4] A sufficient condition for orthogonality is that (a) the number of machines m is odd, (b) the number of jobs n is divisible by m and (c) $\left(\frac{n}{m}\right)$ is an odd number.

Can we find $G = (N, M, s, \Re_+) \in \mathbf{G}$ with $s_1 \neq s_2$ for which $z(s)$ is orthogonal to α_n? The answer to this question is yes. Consider the sub-class

[4] Here $[x]_+$ is the smallest integer greater than or equal to x.

of multiple non-identical machine scheduling problems $\bar{\mathbf{G}} \subset \mathbf{G}$ such that $\forall$ $G^* = (N, M, s, \Re_+) \in \bar{\mathbf{G}}$, (A) n is an odd integer greater than one and (B) $s = (s_1, \ldots, s_m)$ is such that $s_1 = \frac{2}{n}, s_2 = 1 = \ldots = s_m = 1$.

Proposition 2 For all $G^* = (N, M, s, \Re_+) \in \bar{\mathbf{G}}$, $\alpha_n.z(s) = 0$.

Proof: For any $G^* = (N, M, s, \Re_+) \in \bar{\mathbf{G}}$, $z(s) = (z(1), \ldots, z(n))$ is such that

$$z(k) = \begin{cases} \frac{2k}{n} & \text{if } k \in \{1, \ldots, \frac{n-1}{2}\} \\ 1 & \text{if } k = \frac{n+1}{2} \\ \frac{2(k-1)}{n} & \text{if } k \in \{\frac{n+3}{2}, \ldots, n\} \end{cases}$$

Observe that for all $k \in \{1, \ldots, \frac{n-1}{2}\}$, (i) $\frac{z(k)+z(n+1-k)}{2} = z\left(\frac{n+1}{2}\right) = 1$ and (ii) $\alpha(k) = \alpha(n+1-k)$. Condition (ii) holds since n is an odd integer greater than one. Moreover, (iii) $\sum_{k=1}^{n} \alpha(k) = 0$. Using conditions (i), (ii) and (iii) we get $\alpha_n.z(s) = z\left(\frac{n+1}{2}\right) \sum_{k=1}^{n} \alpha(k) = 0$. $\square$

Proposition 2 establishes that the class of multiple non-identical machine scheduling problems with a mechanism satisfying EFF, DSIC and BB is non-empty.

4 Conclusion

In this paper we have identified the orthogonality condition that determines the unique class of multiple non-identical machine scheduling problems for which balanced VCG mechanism exists. One open question is: what happens if we relax the assumption that all jobs have identical processing time in any given machine. The information revelation mechanisms, studied in this paper, are dominant strategy mechanisms. Dominant strategy mechanisms are immune only to unilateral deviations. One important open question is whether there exists a class of multiple non-identical machine scheduling problems for which one can find mechanisms that satisfy efficiency of decision, budget balance along with some form of immunity against group deviations.

References

1. Brucker P (1995) Scheduling Algorithms, Springer-Verlag, New York.
2. Clarke EH (1971) Public Choice 11:17-33
3. Dudek RA, Panwalkar S, Smith M (1975) Management Science 21:544-549
4. Dudek RA, Teuton Jr. OF (1964) Operations Research 12:471-497
5. Green J, Laffont JJ (1979) Incentives in Public Decision Making. North Holland Publication, Amsterdam.
6. Groves T (1973) Econometrica 41:617-631
7. Groves T, Ledyard JO (1977) Public Choice 29:107-124
8. Groves T, Loeb M (1975) Journal of Public Economics 4:211-226

9. Holmström B (1979) Econometrica 47:1137-1144
10. Hurwicz L, Walker M (1990) Econometrica 58:683-704
11. Krabbenborg M, Potters J, van den Nouweland A (1992) European Journal of Operational Research 62:38-46
12. Liu L, Tian G (1999) Review of Economic Design 4:205-218
13. Mitra M (2001) Economic Theory 17:277-305
14. Mitra M (2002) Review of Economic Design 7:75-91
15. Mitra M (2005) European Journal of Operational Research 165:251-266
16. Suijs J (1996) Economic Design 2:193-209
17. Tian G (1996) Economics Letters 53:17-24
18. Vickrey W (1961) Journal of Finance 16:8-37
19. Walker M (1980) Econometrica 48:1521-1540.

Modelling Financial Time Series

P. Manimaran[1], J.C. Parikh[1], P.K. Panigrahi[1] S. Basu[2], C.M. Kishtawal[2] and M.B. Porecha[1]

[1] Physical Research Laboratory, Ahmedabad 380 009, India. parikh@prl.res.in
[2] Space Applications Centre, Ahmedabad 380 015, India.

Summary. Financial time series, in general, exhibit average behaviour at "long" time scales and stochastic behaviour at 'short" time scales. As in statistical physics, the two have to be modelled using different approaches – deterministic for trends and probabilistic for fluctuations about the trend. In this talk, we will describe a new wavelet based approach to separate the trend from the fluctuations in a time series. A deterministic (non-linear regression) model is then constructed for the trend using genetic algorithm. We thereby obtain an explicit analytic model to describe dynamics of the trend. Further the model is used to make predictions of the trend. We also study statistical and scaling properties of the fluctuations. The fluctuations have non-Gaussian probability distribution function and show multi-scaling behaviour. Thus, our work results in a comprehensive model of trends and fluctuations of a financial time series.

1 Introduction

In this presentation, I shall report on the work we have done to study financial time series.

If one observes time series of stock prices, exchange rate of currencies or commodity prices one finds that they have some common characteristic features. They exhibit smooth mean behaviour (cyclic trends) at long time scales and fluctuations about the smooth trend at much shorter time scales. As illustrative examples, values of S&P CNX Nifty index of NSE at 30 second interval and daily interval are shown in Figs. 1 and 2 respectively.

Note also that the time series are not stationary.

In order to model [2] series having these features, we follow the standard approach in statistical physics. More precisely, we consider mean (trend) to have deterministic behaviour and fluctuations to have stochastic behaviour. Accordingly, we model them by quite different methods – smooth trend by deterministic dynamics and fluctuations by probabilistic methods.

In view of this, we first separate mean behaviour from fluctuations in the data. Once this is done, we can decompose a time-series $\{X(i), i = 1, ...N\}$

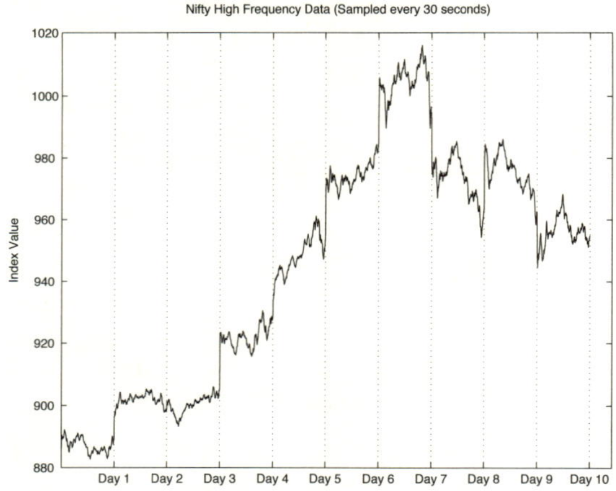

Fig. 1. Nifty high frequency data sampled every 30 seconds for 10 days in January, 1999. (Dates: January 1,4,5,6,7,8,11,12,13,14)

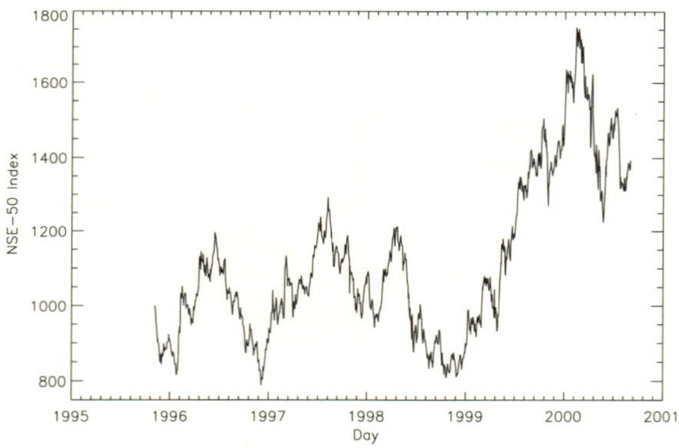

Fig. 2. The daily closing price of S&P CNX Nifty index for the period November 3, 1995-April 11, 2000

as

$$X(i) = X_m(i) + X_f(i) \quad (i = 1, 2, ...N) \tag{1}$$

where $X_m(i)$ is the mean component and $X_f(i)$ the fluctuating component. To carry out this separation we have proposed [3] and used discrete wavelet transforms (DWT) [4] a method well suited for non-stationary data. Section 2 describes in brief some basic ideas of DWT and how it is used to determine the smooth component $\{X_m(i)\}$. The fluctuating component $\{X_f(i)\}$ is obtained by subtracting $\{X_m(i)\}$ from $X(i)$ (see Eq.(1)). We have also compared [3] our procedure with other methods [5] of extracting fluctuations, by studying scaling properties of a computer generated non-stationary series.

In Section 3, we model the smooth series $\{X_m(i);\ i = 1, 2, ...N\}$ in the form of a general non-linear regression equation. The form of the "best" non-linear function is determined by fitting the data and use of Genetic Algorithm [6]. Out of sample predictions are made using the model.

Section 4 contains a discussion of statistical properties of the fluctuations $\{X_f(i);\ i = 1, 2, ...N\}$. Finally, a summary of our work and some concluding remarks are given in Section 5.

2 Discrete wavelets – separation of trend from fluctuations

Wavelet transforms [4] decompose a signal $S(t)$ into components that simultaneously provide information about its local time and frequency characteristics. Therefore, they are very useful for studying non-stationary signals. We use discrete wavelets in our work because one can obtain a complete orthonormal basis set of strictly finite size. This gives significant mathematical and computational advantages. In the present work, we have used Coiflet-2 (Cf-2) wavelets [4].

In wavelets one starts from basis functions, father wavelet $\phi(t)$ (scaling function) and mother wavelet $\psi(t)$ [4]. These functions obey, $\int \phi(t)dt = A$ and $\int \psi(t)dt = 0$, where A is a constant. Scaling and translation of wavelets lead to $\phi_k = \phi(t-k)$ and $\psi_{j,k} = 2^{j/2}\psi\left(2^j t - k\right)$, which satisfy the orthogonality conditions: $\langle \phi_k \mid \phi_{k'} \rangle = \delta_{kk'}$, $\langle \phi_k \mid \psi_{j'k'} \rangle = 0$ and $\langle \psi_{j,k} \mid \psi_{j',k'} \rangle = \delta_{j,j'}\delta_{k,k'}$. Any signal belonging to $L^2(R)$ can be written in the form,

$$S(t) = \sum_{k=-\infty}^{\infty} c_k \phi_k(t) + \sum_{k=-\infty}^{\infty} \sum_{j=0}^{\infty} d_{j,k} \psi_{j,k}(t) \tag{2}$$

where $c_k^{\prime}$s are the low-pass coefficients and $d_{j,k}^{\prime}$s are the high-pass coefficients. We next describe the way we get the smooth trend series starting from the data series.

A forward Cf-2 transformation is first applied to the time series $\{X_i;\ i = 1, 2, ...N\}$. This gives N wavelet coefficients. Of these $\left(\frac{N}{2}\right)$ are low pass coefficients that describe the average behaviour locally and the other half $\left(\frac{N}{2}\right)$ are high pass coefficients corresponding to local fluctuations. In order to obtain the mean (smooth) series, we set the high pass coefficients to zero and then an inverse Cf-2 transformation is carried out. This results in "level one" time series of trend in which fluctuations at the smallest time scale have been filtered out. One can repeat the entire process on "level one" series to filter out fluctuations at the next higher time scale to get a "level two" trend series and so on. Fig. 3 shows the actual $\{X(i)\}$ and (Cf-2, level 4) trend series $\{X_m(i)\}$ of the NASDAQ composite index. Having determined $\{X_m(i)\}$ the fluctuation series $\{X_f(i)\}$ is obtained using Eq. (1).

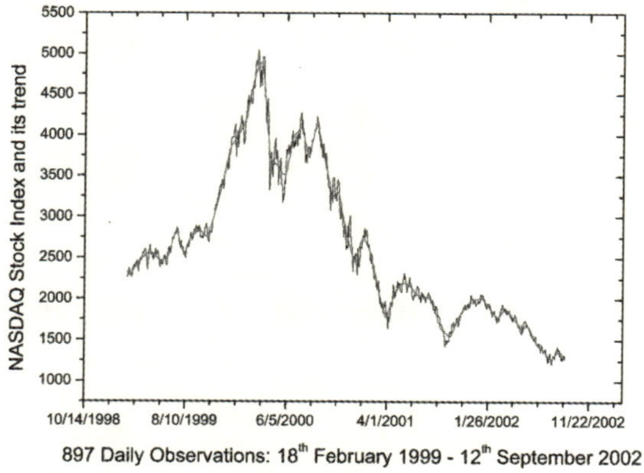

Fig. 3.

It is important to test the validity of our method. For this purpose, we have studied [3] scaling behaviour of fluctuations of computer generated Binomial Multi-Fractal (BMF) series [7].

The qth order fluctuation function $F_q(s)$ [5] is obtained by squaring and averaging fluctuations over M_s forward and M_s backward segments:

$$F_q(s) \equiv \left\{ \frac{1}{2M_s} \sum_{b=1}^{2M_s} \left[F^2(b,s)\right]^{q/2} \right\}^{1/q} \tag{3}$$

Here, 'q' is the order of moment that takes real values and s is the scale. The above procedure is repeated for variable window sizes for different values of q (except $q = 0$). The scaling behaviour is obtained by analyzing the fluctuation function,

$$F_q(s) \sim s^{h(q)} \tag{4}$$

in a logarithmic scale for each value of q. If the order $q = 0$, direct evaluation of Eq. (3) leads to divergence of the scaling exponent. In that case, logarithmic averaging has to be employed to find the fluctuation function:

$$F_q(s) \equiv \exp\left\{ \frac{1}{4M_s} \sum_{b=1}^{2M_s} \ell n \left[F^2(b,s)\right] \right\} \tag{5}$$

As is well-known, if the time series is mono-fractal, the $h(q)$ values are independent of q. For multi-fractal time series, $h(q)$ values depend on q.

In Table 1, we compare exactly known analytical values of the scaling exponent $h_{ex}(q)$ of BMF with the ones obtained by our wavelet based method ($h_w(q)$) and those of the earlier [5] method ($h_p(q)$) using local polynomial fit. Note that the wavelet based approach gives excellent results.

Table 1. The $h(q)$ values of binomial multi-fractal series (BMF) computed analytically ($h_{ex}(q)$) through MF-DFA ($h_p(q)$) and wavelet ($h_w(q)$) approach, Db-8 wavelet has been used.

q	$h_{ex}(q)$	$h_p(q)$	$h_w(q)$
-10	1.9000	1.9304	1.8991
-9	1.8889	1.9184	1.8879
-8	1.8750	1.9032	1.8740
-7	1.8572	1.8837	1.8560
-6	1.8337	1.8576	1.8319
-5	1.8012	1.8210	1.7981
-4	1.7544	1.7663	1.7473
-3	1.6842	1.6783	1.6641
-2	1.5760	1.5397	1.5218
-1	1.4150	1.3939	1.3828
0	0	1.2030	1.2163
1	1.0000	0.9934	1.0091
2	0.8390	0.8312	0.8453
3	0.7309	0.7234	0.7359
4	0.6606	0.6538	0.6649
5	0.6139	0.6075	0.6177
6	0.5814	0.5753	0.5848
7	0.5578	0.5519	0.5610
8	0.5400	0.5343	0.5430
9	0.5261	0.5205	0.5290
10	0.5150	0.5095	0.5178

We have also evaluated the scaling exponent $h(q)$ for the NASDAQ data. This is shown in Fig. 4 where $h(q)$ is a non-linear function of q indicating that the fluctuations have multi-fractal nature.

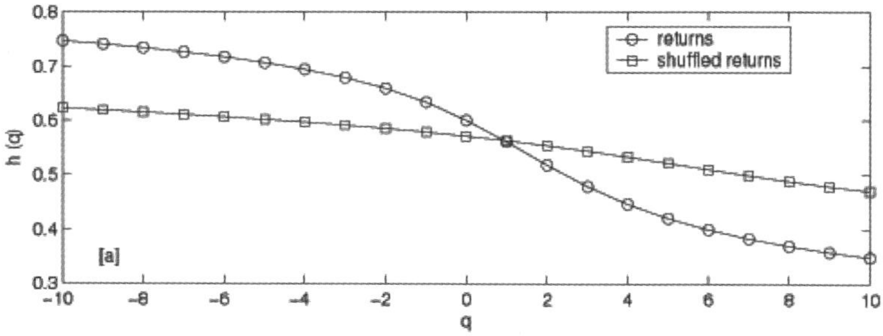

Fig. 4. Scaling exponent for the NASDAQ fluctuations

3 Model of the trend series

We now construct a deterministic model for the smooth series $\{X_m(i)\}$ shown in Fig. 3. For convenience of notation, we define

$$y(i) \equiv X_m(i) \quad (i = 1, 2, ...N) \tag{6}$$

Further, we assume that $y(i)$ is a non-linear function of d previous values of the variable y. More precisely, we have

$$y(i) = F((y(i-1), y(i-2), ...y(i-d)) \quad (i = d+1, d+2, ...N) \tag{7}$$

where F is as an unknown non-linear function and the number of lags d is also not known. It is worth pointing out that, in the study of chaotic time series [8], using the method of time delays to reconstruct the state space, d is actually the embedding dimension of the attractor. We determine its value from false nearest neighbour analysis [8]. For the smoothened NASDAQ series of Fig. 3, the embedding dimension $d = 4$. Therefore,

$$y(i) = F(y(i-1),\ y(i-2)....y(i-4)) \quad (i = 5, ...N) \tag{8}$$

We now use genetic algorithm (GA) [6] to obtain the function F that best represents the deterministic map of the trend series. Following the GA approach of ref. [6] we begin by constructing 200 randomly constructed strings. These strings contain random sequences of the four basic arithmetic symbols $(+, -, \times, \div)$, the values of variable y at earlier times and real number constants. This choice implies that in the GA framework, the search for function F will be restricted to the form of ratio of polynomials – in a way similar to Padé approximation. The equation strings were tested on a training set consisting of first 700 data points of the trend series. A measure of fitness was defined to evaluate performance of each string. These strings are then ordered in decreasing value of fitness and combined pairwise, beginning with the fittest. We retain 50 such pairs which in turn reproduce so that each combined string has 4 offsprings. The first two are identical to the two parents and the remaining two are formed by interchanging parts. A small percentage of the elements in strings are mutated. At this stage fitness is again computed and the entire cycle of ordering, combing, reproducing and mutation of strings is repeated. This is continued for 5000 iterations.

The map that results for our data after carrying out this procedure has the form

$$y(i) = \frac{y(i-4) * (y(i-1))^2}{(y(i-3))^3} \quad (i = 5, ...700) \tag{9}$$

The fitness was 0.9.

Clearly, if the map is a good representation of the dynamics, we ought to have good out of sample predictions. These are shown in Fig. 5.

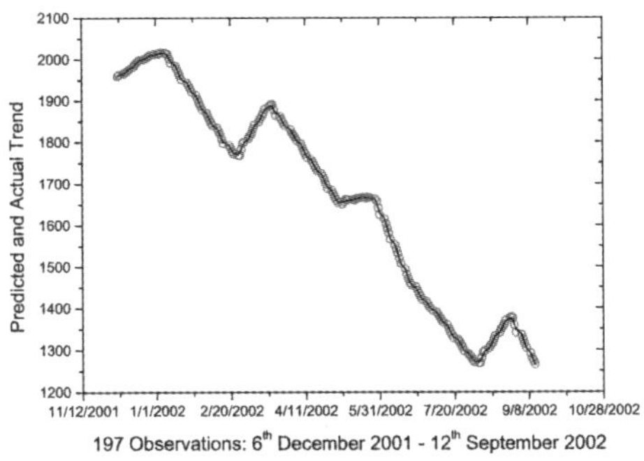

Fig. 5.

We get very promising results – the mean error is $< 0.1\%$ and the sign mismatches are in only about 5% of the predictions. Similar results were also obtained for the BSE Sensex.

It is worth stressing that these are one time step ahead predictions. If the map (Eq.(9)) is iteratively used to make dynamic predictions, then the error grows very quickly. This suggests that the long terms dynamics is not captured by the map.

4 Statistical properties of fluctuations

These properties have been extensively studied and reported in literature (e.g., see refs. [1]-[2]). All the same, for the sake of completeness, we show in the figures below:

(i) Probability distribution function (PDF) of returns (Fig. 6)
(ii) Auto-correlation function of returns (Fig. 7)
(iii)Auto-correlation function of absolute value of returns (Fig. 8)

Fig. 6 shows that the PDF is not Gaussian – it has skewness = -0.07 and kurtosis = 8.27. Note that for a Gaussian PDF these reduced cumulants are zero. Figs. 7 and 8 together show that the binary correlations go to zero quickly but the "volatility" correlations persist for a long time. This means that fluctuations are not independent.

As mentioned earlier, these fluctuation properties are well-known to occur in financial markets [1,2].

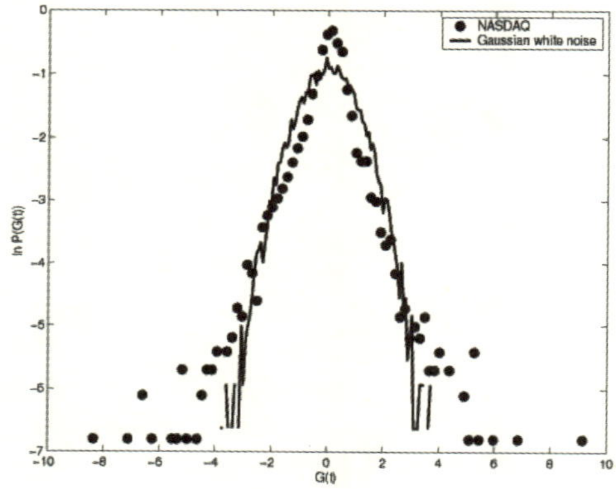

Fig. 6. Probability distribution function (PDF) of returns

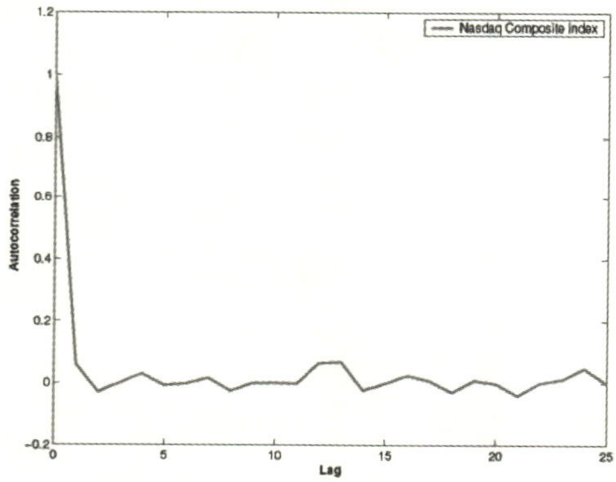

Fig. 7. Auto-correlation function of returns

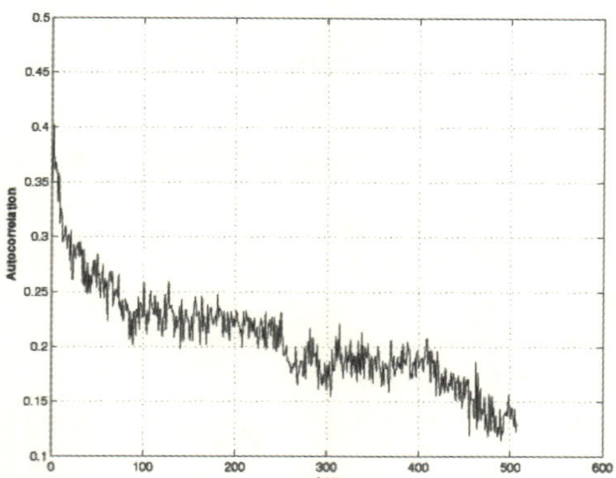

Fig. 8. Auto-correlation function of absolute value of returns

5 Summary and future work

Observation of financial market data suggests that the dynamics of the market may be viewed as a superposition of long term deterministic trend and short scale random fluctuations.

We have used discrete wavelet transforms to separate the two parts and developed suitable models.

The trend has been modelled in the form of a non-linear auto-regressive map. An explicit analytic functional form for the map is found using GA. It gives an excellent fit to the data set and makes very reliable out of sample single step predictions. Multi-step predictions are not good.

Statistical and scaling properties of fluctuations were determined and as expected were consistent with earlier findings [1,2].

Regarding further work in this direction, a major challenge is to apply GA methodology to arrive at a model with capability of making long term predictions of the trend series.

In conclusion, we have thus made a beginning towards simultaneously modelling trend and fluctuations of a time series.

References

1. Mandelboot BB (1997) Fractals and Scaling in Finance, Springer-Verlag, New York; Bouchaud JP, Potters M (2000) Theory of Financial Risk, Cambridge University Press, Cambridge; Mantegna RN, Stanley HE (2000) An Introduction to Econophysics, Cambridge University Press, Cambridge
2. Parikh JC (2003) Stochastic Processes and Financial Markets, Narosa, New Delhi
3. Manimaran P, Panigrahi PK, Parikh JC (2005) Phys. Rev. E72:046120-
4. Daubechies I (1992) Ten Lectures on Wavelets, SIAM, Philadelphia
5. Kantelhardt JW, Zschiegner SA, Koscielny-Bunde E, Havlin S, Bunde A, Stanley HE (2002) Physica A 316:87
6. Szpiro GG (1997) Phys. Rev. E 55:2557
7. Feder J (1998) Fractals, Plenum Press, New York
8. Abarbanel H, Brown R, Sidorovich J, Tsimring L (1993) Rev. Mod. Phys. 65:1331.

Random Matrix Approach to Fluctuations and Scaling in Complex Systems

M. S. Santhanam

Physical Research Laboratory, Navrangpura, Ahmedabad 380 009, India.

1 Introduction

The study of fluctuations, self-similarity and scaling in physical and socio-economic sciences in the last several years has brought in new insights and new ideas for modelling them. For instance, one of the important empirical results of the market dynamics is that the probability distribution of price returns r in a typical market displays a power-law, i.e, $P(|r| > x) \sim r^{-\alpha}$, where $\alpha \sim 3.0$ [1]. In fact, this "inverse cube law" is known to hold good for volume of stocks traded in stock exchanges, though the exponent in this case is $\alpha \sim 1.5$ [1]. Similar power laws appear for the cumulative frequency distribution of earth quake magnitudes, often called the Gutenberg-Richter relation [2]. Infect, anything from size distribution of cities and wealth distributions, display power law. These apparently universal power laws pertain to the distribution of actual values taken by some quantity of interest, say, a stock market index and these distributions reveal scaling with certain parameters.

On the other hand, we also know that the fluctuations of many processes are correlated and are self-similar. It has been shown that the fluctuations in the indices of major stock markets are long range correlated [3], physically implying that the correlations in the *fluctuations* do not decay fast enough and the system possesses long memory. This property is also true of EEG series, temperature anomalies worldwide and even use of words in English language by authors such as Shakespeare and Charles Dickens [4]. This arises from very specific kinds of correlation existing among the fluctuations after the trends are removed. Here we shall seek to find analogues for such time series correlations in the fluctuations of level spectra of quantum systems. This necessarily involves studying the correlations among the quantum levels for which an appropriate tool is the random matrix theory (RMT). Hence, we will attempt to map the properties of the ensembles of random matrix theory with those of correlated time series. At this point, we emphasise again that we are attempting find analogues for correlations and scaling in *univariate* time series within the context of level spectra from random matrices.

Note that RMT has already been applied to understand non-trivial correlations in multivariate time series, such as the time series from N different stocks or sea level pressure from N different locations [5]. We shall first briefly review the latter results. Consider a multivariate time series of length L for M, say, stocks operating in a stock market. If $\mathbf{Z}$ is the data matrix with elements r_{tq} that denote the return of the qth stock at tth instant, then the correlation matrix is given by,

$$\mathbf{C} = \frac{1}{L}\mathbf{Z}^{\mathbf{T}}\mathbf{Z} \tag{1}$$

By construction, this is a real symmetric matrix. Now one could consider an equivalent random matrix $\mathbf{R}$ whose entries are taken from a standard normal distribution. Then, the random correlation matrix is, $\mathbf{C_R} = (\mathbf{1/L})\mathbf{R^T R}$. By comparing the statistical properties of the spectra of $\mathbf{C}$ and $\mathbf{C_R}$, it has been shown that $\mathbf{C}$, for most part, can be modelled as a random matrix taken from an appropriate ensemble of random matrix theory [5]. The deviations from RMT prescribed properties can be identified with specific information carried by the multivariate time series. This approach has been applied to stock markets, atmospheric data, correlations in internet traffic and EEG recordings [6]. Recently, delay correlation matrix has also been considered within the ambit of RMT and is shown to display new information about correlations with a delay [7].

In this work, we will develop methods to treat univariate time series in the framework of RMT. With this introduction, we plunge into the results. In the next section, we will develop the formalism to treat the quantum levels as a time series. Subsequently, we show that the spectral transition in RMT can be mapped to an anti-persistent time series with Hurst exponent $0 < H \leq 1/2$. Finally, we discuss some applications of these results.

2 Quantum levels as time series: formalism

Consider a quantum system with $n+1$ discrete levels $E_1, E_2, E_3, ...E_{n+1}$. For the present purpose, in an approximate sense, it is sufficient to think of this set of discrete levels to be the eigenvalues of a real symmetric matrix. The integrated level density, a counting function that gives the number of levels below a given level E, is given by,

$$N(E) = \sum_i \Theta(E - E_i) = N_{avg}(E) + N_{fl}(E). \tag{2}$$

In order to remove the trends in integrated level density, we 'unfold' the spectrum. We define the unfolded levels as, $\lambda_i = N_{avg}(E_i)$ such that the mean level density is equal to unity. All further statistical analysis is performed on the sequence of levels $\{\lambda\}$. Now, the spacing between adjacent level is

$s_i = \lambda_{i+1} - \lambda_i$. One of the celebrated results of random matrix theory is the spacing distribution given by [8],

$$P_{GOE}(s) = \frac{\pi}{2} s \exp\left(-\frac{\pi}{4} s^2\right), \tag{3}$$

where GOE indicates Gaussian orthogonal ensemble of random matrix theory, a collection of real symmetric matrices with Gaussian distributed random numbers with specified mean and variance [8]. This is the so-called level repulsion regime, i.e, probability of degeneracies is very low. It is also established that for a collection of real diagonal matrices the spacing distribution of its unfolded levels is Poissonian, $P(s) = \exp(-s)$, i.e, the spacings are uncorrelated. In RMT language, this belongs to the so-called Gaussian diagonal ensemble (GDE), an ensemble of diagonal matrices with uniformly distributed random numbers. For a rigorous definition of various ensembles of RMT, we refer the reader to the book by Mehta [8]. As an important aside we also note that RMT has important applications in nuclear physics, quantum chaos, mesoscopic and other condensed matter systems [9]. Now, we define the time series of levels by,

$$\delta_m = \sum_i^m (s_i - \langle s \rangle) \equiv -N_{fl}(E). \tag{4}$$

Now, δ_m is formally analogous to a time series [10], with m serving as the discrete time index. , the time series look distinctly different for the cases when the levels are from GDE class matrix as opposed to a GOE type matrix. In Fig 1, we show δ_m calculated for levels from GDE and GOE type matrices. Notice that δ_m for GDE case (gray line) resembles a random walk time series whereas the GOE levels (black line) seems to display short range correlation.

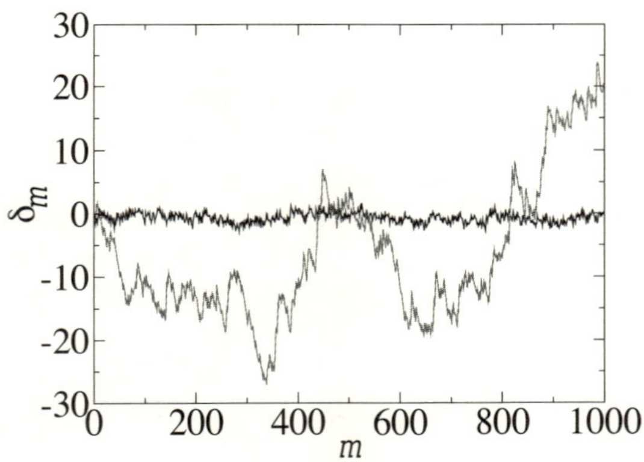

Fig. 1. The 'time series' of levels δ_m plotted as a function of m. The black line represents levels taken from GOE class of matrices. The Gray line corresponds to levels taken from GDE class of matrices.

2.1 Power spectrum

Since we have interpreted the levels of a quantum system as a time series, we can use tools of time series analysis like the power spectrum to study them. Now we state the central result at this point. The ensemble averaged power spectrum $\langle S(f) \rangle$ of the time series of levels is given by [10],

$$\begin{aligned} \langle S(f) \rangle &\propto 1/f^2 && \text{(GDE)} \\ &\propto 1/f && \text{(GOE)} \\ &\propto 1/f^\gamma && \text{(GDE to GOE transition)} \end{aligned} \tag{5}$$

Thus, levels from GOE type matrices display $1/f$ noise whereas the ones from GDE type matrices display $1/f^2$ noise. The transition from GOE to GDE displays a general $1/f^\gamma$ noise where $1 < \gamma < 2$. In Fig 2, we show the power spectrum for levels from GDE and GOE calculated from an ensemble of 50 matrices of order 5000. Note that the agreement between Eq 5 and the numerical simulations is quite good. This result also holds good for quantum systems whose classical dynamics makes a transition from regularity to chaos [11].

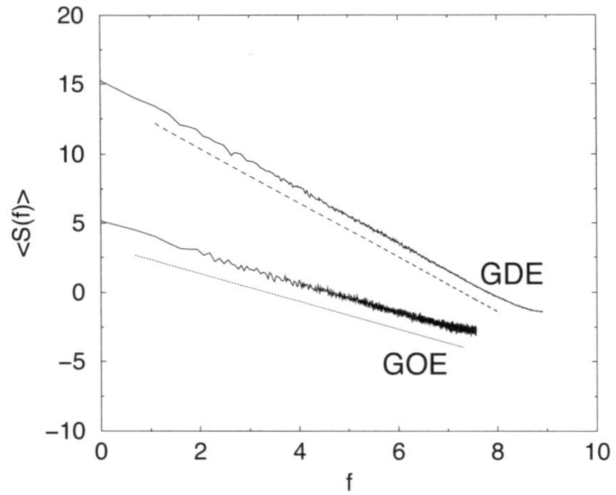

Fig. 2. Ensemble averaged power spectrum for levels of GDE and GOE. The broken lines are the of best fit iones obtained by linear regression. The slopes are -1.96 (GDE) and -1.02 (GOE). This is in agreement with Eq 5.

3 RMT and time series models

In this section, we will map our results in Eq 5 to time series models. From the functional forms of the power spectrum it is straightforward to figure out that $1/f^2$ type power spectrum must come from a random walk like time series. Similarly, a very particular kind of short range correlations among the levels leads to $1/f$ type power spectrum. Hence, we are looking at a fractional Brownian motion model to serve as the interpolating time series between these two limits. It is well known that the fractional Brownian motion (fBm) series $g(t)$ is a self-affine statistical fractal, meaning that it displays a scaling relation of the form,

$$g(Ht) \approx t^{-H} g(t)$$

where the symbol $\approx$ indicates statistical equality and the scaling exponent $0 < H < 1$ is called the Hausdorff measure. It is also called the Hurst exponent. For the fBm case, we can write the variance as, $\mathrm{var}(g) \sim t^{2H}$. The standard random walk corresponds to $H = 1/2$. Since we know the general relation between H and γ to be $\gamma = 2H + 1$, we can relate the statistical properties of the time series of levels to a scaling exponent H for the time series. Hence, the entire range of $1 < \gamma < 2$ can be mapped to $0 < H \leq 1/2$. Thus, the time series of levels, making a transition from GOE to GDE, are an anti-persistent series. This anti-persistence implies that the values of time series at subsequent intervals are anti-correlated. This is also inferred from the fact at $H = 1/2$ the time series is much more smoother compared to the one at $H = 0$, as seen in Fig 1.

How do we determine the value of H ? This is done by performing an R/S analysis or detrended fluctuation analysis (DFA). As a next step, we relate the DFA fluctuation function to $\Delta_3(s)$ statistics widely used in random matrix theory. Then, using the results from RMT we obtain analytically the DFA fluctuation function and the values of H for time series of levels drawn from GDE and GOE.

4 Detrended fluctuation analysis and RMT

In the last one decade DFA has emerged as an important technique to study scaling and long range correlations in non-stationary time series [12]. DFA is based on the idea that if the time series contains nonstationarities, then the variance of the fluctuations can be studied by successively detrending using linear, quadratic, cubic ... higher order polynomials in a piecewise manner. The entire time series is divided into segments of length s and in each of them detrending is applied and variance $F^2(s)$ about the trend is computed and averaged over all the segments. This exercise is repeated for various values of s. For a self-affine fractal, we have,

$$F(s) \propto s^H \qquad (0 < H < 1) \tag{6}$$

We will not describe it further here since DFA has been extensively studied in the literature [12]. In order to obtain analytical estimate for the value of H, we appeal to the $\Delta_3(s)$ statistic widely studied in RMT to quantify the rigidity of spectrum, i.e, given a level, a rigid spectrum will not allow any flexibility in the placement of the next level. $\Delta_3(s)$ is a measure of this flexibility or conversely the rigidity. It is defined by [8],

$$\Delta_3(\bar{\lambda}, s) = \frac{1}{2s} \min_{a,b} \int_{\bar{\lambda}-s}^{\bar{\lambda}+s} [n(\lambda) - a\lambda - b]^2 d\lambda \tag{7}$$

where $n(\lambda) = n_{av}(\lambda) + n_{osc}(\lambda)$ is the integrated level density for the unfolded spectrum $\{\lambda\}$ and a and b are the parameters from linear regression and $\bar{\lambda}$ is any reference level. Now, $\Delta_3(s)$ is an average over various choices of $\bar{\lambda}$. It can be shown that [13],

$$\langle F_1(s) \rangle \sim \langle \Delta_3(s)^{1/2} \rangle \tag{8}$$

where the F_1 indicates DFA fluctuation function with linear detrending. Now, the RMT averages for $\langle \Delta_3(s) \rangle$ are well known from RMT literature. We substitute them in Eq 8 and immediately we obtain results valid for correlated time series. First, we consider GDE case. The RMT average is [8], $\langle \Delta_3(s) \rangle = s/15$. Substituting in Eq 8, we obtain

$$\langle F_1(s) \rangle \sim s^H \sim (s/15)^{1/2} \tag{9}$$

We have just reproduced the known result that for the random walk, $H = 1/2$.

Next, we consider the GOE case. Firstly we note that with a little more involved calculation it can be shown that GOE levels correspond to $H = 0$ [13]. The required result from RMT is [8], $\langle \Delta_3(s) \rangle \sim \ln s/\pi^2$. Substituting in Eq 8, we obtain,

$$\langle F_1(s) \rangle = (\ln s/\pi^2)^{1/2}. \tag{10}$$

This shows that the square of the fluctuation function F_1^2 for the GOE case ($H = 0$) has a logarithmic behaviour. , as our simulations in Fig 3 demonstrate, logarithmic behaviour sets in even as $H \to 0$. This result is not all that well known. There have been experimental results on fluctuations in surface roughness of kinetic roughening models which show that at $H = 0$ the fluctuations have a log-like behaviour [14]. Here we obtain it by considering the connection between RMT and detrended fluctuation analysis.

5 Application to real time series

Firstly, we illustrate our results in Eq 10 using a simulated time series. In Fig 3, we show the DFA performed on a synthetically generated time series with $H = 0.1, 0.2, 0.3$. This can be done by the method presented Ref [15]. In all the three cases, the log behaviour is clearly seen as evident from the straightlines in semi-log plot shown alongside.

Since the RMT analogy presented here restricts the possible values of H to less than $1/2$, we need to look at time series within this constraint. This means that the possible candidates must exhibit anti-persistent behaviour. Most real life time series exhibit persistence, i.e, subsequent elements of the time series are correlated. However, there are several examples of anti-persistent time series. For instance, electricity spot prices in Scandinavia [16], southern oscillation index [17], and foreign exchange rates of some (currently defunct) currencies [18] display anti-persistence. We illustrate our results with the daily foreign exchange rates for Belgium Francs - Deutsche Marks during 1991 - 1994 [19]. In Fig 4 we show the DFA fluctuation function for the BEF-DEM which displays logarithmic behaviour expected from Eq 10. , a semi-log plot (not shown here) reveals a good straight line confirming the logarithmic behaviour.

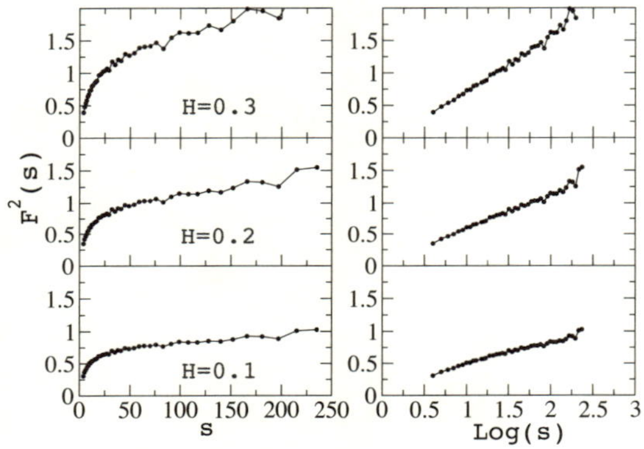

Fig. 3. The DFA fluctuation function $F_1^2(s)$ plotted for H =0.1, 0.2 and 0.3. Notice that as $H \to 0$, DFA function displays log behaviour as predicted by Eq 10. This is clearly seen as straight lines in the semi-log plot shown along side.

The result in Eq 10 has an important application in the identification of cross-over behaviour in fluctuation functions. In anti persistent series, often a logarithmic behaviour could be mistaken for two or more power law functions. This leads to identification of spurious cross-overs where none exist. Hence, while dealing with anti-persistent series it is important to check for logarithmic behaviour before identifying cross-overs in time series data.

6 Conclusions

We started by reinterpreting the levels of random matrices as a time series. By considering the power spectrum of such a time series of levels, we are able map the level spectrum described by Gaussian ensembles of random matrix theory to a self-affine time series with Hausdorff measure in the range $0 < H < 1/2$. Then, we show that the fluctuation function of detrended

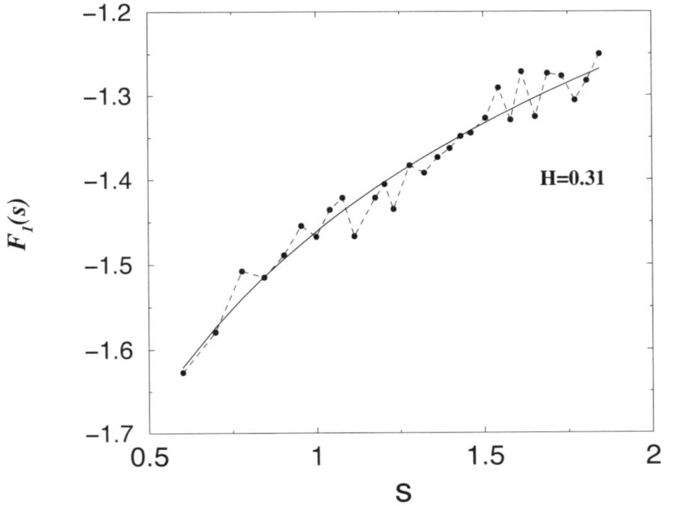

Fig. 4. The DFA fluctuation function $F_1^2(s)$ plotted for Belgium Francs - Deutsch Marks exchange rates during 1991 - 1994 (solid circles). The value of $H = 0.31$. Note the best fit logarithmic function (solid line) which closely follows the trend displayed by the DFA function (solid circles).

fluctuation analysis is equivalent to $\Delta_3(s)$ statistics in RMT. We exploit this connection to analytically obtain the fluctuation function for the time series. We illustrate our results with an example from real life economic time series and discuss its applications in identifying cross-over phenomena in data.

References

1. Gabaix X et al. (2003) Nature 423:267-270.
2. Newman MEJ (2005) Contemporary Physics 46(5):323
3. Liu Y et al. (1999) Phys. Rev. E 60:1390; Yamasaki et al (2005) Proc. Nat. Acad. Sci. 102:9424.
4. Eva Koscielny-Bunde et al. (1998) Phys. Rev. Lett. 81:729-732; Montemurro MA, Pury PA (2002) Fractals 10(4):451-461
5. Plerou V et al. (1999) Phys. Rev. Lett. 83:1471-1474; Laloux L et al. (1999) Phys. Rev. Lett. 83:1467-1470; Santhanam MS, Patra PK (2001) Phys. Rev. E 64:016102
6. Plerou V et al. (2002) Phys. Rev. E 65:066126; Barthelemy M et al. (2002) Phys. Rev. E 66:056110; P. Seba (2003) Phys. Rev. Letts. 91:198104
7. Mayya KBK, Amritkar RE, Santhanam MS (2006) cond-mat/0601279
8. Mehta ML (1995) Random Matrices. Academic Press, New York.
9. Guhr T, Muller-Groeling A, Weidenmuller HA (1998) Phys. Reports 299:189-425

10. Relaño A et al. (2002) Phys. Rev. Lett. 89:244102; Faleiro E et al. (2004) Phys. Rev. Lett. 93:244101
11. Gomez JMG et al. (2005) Phys. Rev. Lett. 94:084101; Santhanam MS, Bandyopadhyay JN (2005) 95:114101
12. Peng C-K et al. (1994) Phys. Rev. E 49:16851689
13. Santhanam MS, Bandyopadhyay JN, Angom D (2006) Phys. Rev. E 73:015201(R)
14. Palasantzas G (1993) Phys. Rev. E 48:1447214478
15. Rangarajan G, Ding M (2000) Phys. Rev. E 61:4991-5001
16. Weron R, Przbylowicz B (2000) Physica A 283:462-468; Weron R (2001) cond-mat/0103621
17. Ausloos M, Ivanova K (2001) Phys. Rev E 63:047201
18. Ausloos M (2000) Physica A 285:48-65
19. The BEF-DEM currency data is obtained online from Pacific Exchange Rate Service, University of British Columbia, Canada.

The Economic Efficiency of Financial Markets

Yougui Wang

Department of Systems Science, School of Management, Beijing Normal University, Beijing, 100875, People's Republic of China. ygwang@bnu.edu.cn

Summary. In this paper, we investigate the economic efficiency of markets and specify its applicability to financial markets. The statistical expressions of supply and demand of a market are formulated in terms of willingness prices. By introducing probability of realized exchange, we also formulate the realized market surplus. It can be proved that only when the market is in equilibrium the realized surplus can reach its maximum value. The market efficiency can be measured by the ratio of realized surplus to its maximum value. For a financial market, the market participants are composed of two groups: producers and speculators. The former brings the surplus into the market and the latter provides liquidity to make them realized.

Key words: Financial Markets, Market Efficiency, Market Equilibrium, Producer, Speculator

1 Introduction

Since the efficient markets hypothesis was put forward in fifties of last century, there has been vigorous debate over the question of whether or not the financial markets are efficient. So far no consensus has been achieved. The people who believe that they are efficient argue that all information about a tradable security is always embedded in the security's price and no one can take advantage of any known information to get an excess return from trading the security. Instead those who are in the opposite viewpoint provide many empirical evidences against this hypothesis. However, most of the debaters refers to the information efficiency [1], only a few refers to the economic efficiency [2–4]. There has not been an explicit definition of economic efficiency for a financial market yet.

According to the efficient markets hypothesis, stock prices will move randomly due to the random arrival of all new information. Many stylized facts recently observed contradict this prediction [5]. Even if the price of a stock indeed moves in a random walk, we still can not say that the market is efficient in economic sense. For the same reason, we also can say nothing as to whether it is efficient or not when the price movement appears with a "fat-tail" return distribution. In order to explore the mechanism behind the stylized facts of financial markets, a variety of models have been developed to mimic the markets [6–9]. Few of them has considered the economic efficiency. But when all efforts were put into the investigations on financial markets, I believe, most of the works have been carried out towards the hope the society could benefit more from the markets. Then the problem as to how to measure the total welfare generated by transactions in the markets arises.

In traditional economics, supply and demand are the core tools for analyzing the markets. This static approach is obviously not suitable to the financial markets due to drastic volatility of prices. This fact triggers researchers to find other approaches. Then in order to understand how these markets perform, nonlinear dynamics was proposed to construct the formation of price, agent-based models were developed to produce a time series of price [8], even minority game model was put forward to catch the key ingredients of financial markets [6]. However, when we try to mimic financial markets in a novel way, the essential element which underlies supply and demand analysis should not be discarded.

In this paper, the concept of willingness price will be introduced which is deemed to be the essential element for market analyzing. Based on this concept, a mathematical description of supply and demand is then presented. The market surplus is also formulated in terms of willingness exchange distribution. It can be classified into expected surplus and realized one. Then the market efficiency is shown to be associated with the realized surplus. When turning to the financial markets, I argue that this approach is the only way to analyze this kind of efficiency. However, the surplus is not brought by all traders who participate in the transactions. Only a part of them who are called producers play this role. Others which are called speculators contribute to the market by providing liquidity and filtering the producers. I believe this perspective will be helpful for modelling the financial markets.

2 Measuring the market economic efficiency

2.1 Willingness price

When a market price is given, a rational trader is often assumed to use benefit-cost analysis to make a decision of whether or not to proceed an exchange at this price. This analysis results in a concept of willingness price. The willingness price of a buyer is defined as the maximum amount that he is willing to pay for one unit of the good. On the other hand, the willingness price of a seller is defined as the minimum amount that he is willing to sell one unit of the good. This sometime is usually called 'reservation price' or 'cost' in economics literature. It is in some sense like 'limit order' in financial markets.

It follows from this definition that the willingness price has impact on the behavior of traders in the market. To decide whether or not conclude a business, every trader must make a comparison between actual market price and his own willingness price. In each transaction, given the actual market price p, if the buyer's willingness price, ν^b, is not less than p, i.e.,

$$p \leq \nu^b, \tag{1}$$

then he buys one unit of the good. Otherwise, he gives up his purchasing. On the other side, if a seller's willingness price, ν^s, is not greater than p, i.e.,

$$p \geq \nu^s, \tag{2}$$

then he sells one, above which he will withdraw the offer of his good. Thereforethe necessary condition for closing a bargain is that the buyer's willingness price must be greater than or equal to the seller's willingness price. And the actual price must lies between the buyer's willingness price and the seller's willingness price.

2.2 Supply and demand functions

We consider a market in which traders are all price-takers and each is allowed to trade only one unit of the good in each transaction, the market price of the good is set exogenously. Different traders, including sellers or buyers, may have different willingness prices. It is reasonable to assume that willingness prices spreads over the domain of $(0, \infty)$. Under continuous assumption, there must be a portion of sellers in the market $F_s(\nu)\, d\nu$ whose willingness prices are between ν and $\nu + d\nu$. In the same way, there must also be a portion of buyers $F_d(\nu)\, d\nu$ whose willingness prices are between ν and $\nu + d\nu$. $F_s(\nu)$ and $F_d(\nu)$ are distribution functions for sellers and buyers respectively.

From the exchange conditions specified by equations (1) and (2), given the market price p, the quantity supplied is the portion whose willingness prices are not greater than the actual price, while the quantity demanded is the portion whose willingness prices are not less than actual price. Thus the supply and demand functions $Q_s(p)$ and $Q_d(p)$ respectively are given by,

$$Q_s(p) = \int_0^p F_s(\nu)\, d\nu, \tag{3}$$

and

$$Q_d(p) = \int_p^\infty F_d(\nu)\, d\nu. \tag{4}$$

From (3) and (4) we can see that supply and demand are directly determined by the willing exchange distributions over the willingness price. If these distribution functions are nonnegative and independent of the market price, then we have

$$\frac{dQ_s(p)}{dp} \geq 0, \tag{5}$$

and

$$\frac{dQ_d(p)}{dp} \leq 0. \tag{6}$$

These results are always called the laws of supply and demand respectively in economic literature, which state that when price gets higher, quantity demanded is reduced and more is supplied. The laws for individual supply and demand have always been derived from the marginal returns law and the marginal utility law [10]. However, difficulties arise when one ask whether the laws still hold for the aggregate level, especially for market demand [11]. From Equations (5) and (6) it is clear that as long as the willing exchange distribute over the willingness price, then the laws of demand and supply hold, no matter what level we are considering.

2.3 Market surplus

Economists usually employ market surplus to calculate the benefits that the traders have gained through exchanges in markets. The surplus of an individual transaction is the difference between the buyer's willingness price and the seller's willingness price. Whatever the closing price is, it must within the two willingness prices. So the surplus must be nonnegative and is independent on the price. The impacts that the price has is only on the distribution of the surplus between the buyer and seller. The part above the price is called the buyer's surplus, the other part below the price is the seller's surplus. Given the actual market price of the good, adding up all the corresponding traders' surplus, we have the total seller surplus and buyer surplus of the market respectively,

$$S_s(p) = \int_0^p F_s(\nu)(p - \nu)\, d\nu, \tag{7}$$

$$S_b(p) = \int_p^\infty F_d(\nu)(\nu - p)\, d\nu. \tag{8}$$

At first glance, it seems that the total market surplus is just the sum of buyer surplus and seller surplus. However, not all of the surplus can be realized in actual transactions. The sum is just the surplus that buyers and sellers expect to obtain through their potential exchange in the market at this price. So we denote it the expected market surplus as the following

$$S_e(p) = \int_0^p F_s(\nu)(p - \nu)\, d\nu + \int_p^\infty F_d(\nu)(\nu - p)\, d\nu. \tag{9}$$

In fact, when the price is set at an appropriate level, so that there is no excess supply or demand, we say that the market is in equilibrium. This price is known as the market-clearing price or the equilibrium price. When the price is greater than the equilibrium price, the market is in surplus; when the actual price is less than the equilibrium price, the market is in shortage. In any one of the two cases, the quantity demanded is not equal to the quantity supplied any more, and there must be some willing exchanges can not be realized. When the market is in surplus, all buyers who are willing to purchase can get what they desired, meanwhile there must be some willing sellers who have no chance to make their deals. When the market is in shortage, all sellers can realize their willing exchanges, but some buyers can not get what they are willing to buy. In order to compute the real surplus that the traders have obtained for all these occasions, we introduce probability of realized exchange which is in the domain of $[0, 1]$. Denote the probabilities of realized exchange for sellers and buyers as $g_s(\nu)$ and $g_d(\nu)$ respectively, we than have an identity given by

$$\int_0^p F_s(\nu)g_s(\nu)\ d\nu \equiv \int_p^{\infty} F_d(\nu)g_d(\nu)\, d\nu. \tag{10}$$

It says that the aggregate realized quantity of demand must be equal to the aggregate realized quantity of supply for any case.

Since only the traders who realized their exchanges can get surplus, total realized surplus that buyers and sellers obtain through their actual exchange in a market can be expressed as

$$S_r(p) = \int_0^p F_s(\nu)g_s(\nu)(p-\nu)\ d\nu + \int_p^{\infty} F_d(\nu)g_d(\nu)(\nu - p)\, d\nu. \tag{11}$$

We now have got a measure of market economic efficiency which can be defined by the ratio of realized surplus to its maximum value. We show illustratively the dependence of market economic efficiency on the actual price in Figure 1. It can be seen that the realized total market surplus reaches its maximum value when the market is in the equilibrium. This conclusion can be proved in a strict mathematical way, which is not presented here due to space limitation.

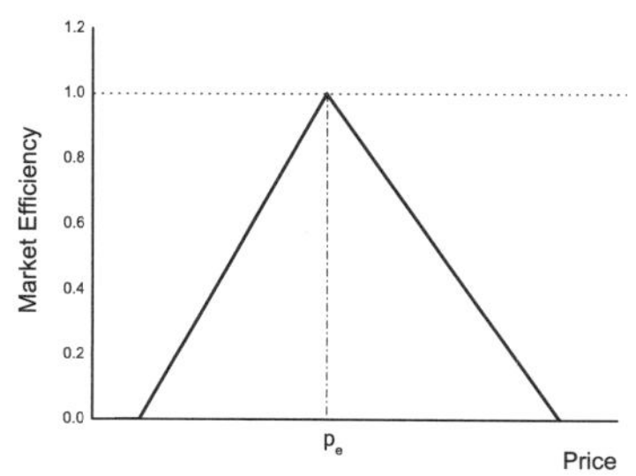

Fig. 1. Illustrative dependence of market economic efficiency on market price

3 Economic efficiency of financial markets

According to the work of Y.-C. Zhang, the traders of financial markets can be divided into two groups: producers and speculators [3]. The producers are those traders whose core economic activities are outside economy. The financial markets are means for them to seek the profit outside the markets. The speculators are those whose aim is to find predictability of the market price and seek profit in financial markets. The producers create economic surplus and bring it into the markets. However, they could not realize the surplus without the liquidity provided by speculators. Thus, the producers and speculators cooperate in the financial markets. The relationship of them is just like symbiosis.

Just like in goods markets, producers in financial markets, including buyers and sellers, have their own willingness prices. When they participate in the trade, they bring surplus into the markets. The realized market surplus is the difference between aggregate willingness prices of producers who have bought securities and aggregate willingness prices of producers who have sold securities. The total surplus will be shared by the producers and speculators as reward to their corresponding efforts. The market economic efficiency can also be measured by the ratio of realized surplus to its maximum value.

Among the factors that govern the market economic efficiency, the price still plays the essential role. The price is mainly determined by the interaction among speculators for they are the majority of the trade participants. For the producers are price-takers, they make their decisions by comparing the actual price and their willingness prices. Thus the price acts likes a filter of producers. If the price the producers face is too high, some sellers with higher cost may be allowed to make exchanges, while some potential buyers may be refused. On the contrary, if the price is too low, potential sellers may be refused and some buyers with lower valuation may be permitted to make exchange. Both cases result in loss due to inappropriate level of price. Now we have a question in our hand, what is the equilibrium price of a financial market which leads to an efficient outcome? Actually, the market is not at a static state but evolves with time, while the traders enter and leave the markets randomly. This process is continuous and can not be separated one by one in terms of periods. Even the value inflow of producers are given, what patterns of price movement can lead to the efficiency? Random walk or one with fat-tail return distribution? These questions call for more efforts in future investigation.

4 Discussion and conclusion

When we turn to understand how a market works, there are two basic and important questions that must be answered. The first one is about how individual buyers and sellers make their choices. It is suggested that besides strategies of speculators, we should consider the willingness prices of producers. Generally, the willingness price of a producer is formed according to the situation he or she faces and may evolve over time. The second one is as to how these traders interact with each other to form the market price. Since the price is determined by supply and demand, the Equations (3) and (4) we get in previous section can be applied in the formation of price.

In summary, the market economic efficiency can be measured by the realized market surplus, which can be expressed in mathematical way based on the concept of the willingness price. In traditional economics, when the market is in equilibrium, it is also efficient. Although this approach can be employed to analyze the efficiency of financial markets, some difficulties should be overcome. The efficiency of financial markets is fundamentally based on the cooperation of producers and speculators.

Acknowledgements

I am grateful to Prof. Honggang Li for his helpful discussions and comments. This work is supported by National Nature Science Foundation of China under Grant Nos. 73071073.

References

1. Malkiel BG (1989) Is the stock market efficient? Science 243:1313–1318
2. Dow J, Gorton G (1997) Stock market efficiency and economic efficiency: Is there a connection. The Journal of Finance 52(3):1087–1129
3. Zhang YC (1999) Toward a theory of marginally efficient markets. Physica A 269:30–44
4. Capocci A, Zhang YC (2001) Market ecology of active and passive investors. Physica A 298:488–498
5. Mantegna R, Stanley HG (2000) Introduction to econophysics. Cambridge University Press, Cambridge
6. Challet D, Zhang YC (1997) Emergence of cooperration and organization in an evolutionary game. Physica A 246:407–418
7. Sato A, takayasu H (1998) Dynamical models of stock market exchange: from microscopic determinism to macroscopic randomness. Physica A 250:231–252
8. Lux T, Marchesi M (1999) Scaling and criticality in a stochastic multi-agent model of a financial market. Nature 397:498–500
9. Levy H, Levy M, Solomon S (2000) Microscopic simulation of financial markets: From investor behavior to market phenomena. Academic Press, London
10. Varian HR (1992) Microeconomic Analysis. Norton, New York
11. Janssen MCW (1993) Microfoundations: A Critical Inquiry. Routledge, New York

Regional Inequality

Abhirup Sarkar

Economic Research Unit, Indian Statistical Institute, 203 B. T. Road, Kolkata 700108, India. abhirup@isical.ac.in

Summary. The purpose of the paper is to analyse the problem of regional inequality and the possibility of inter-regional cooperation. In particular it tries to answer why backward regions of a country might prefer regional autonomy as opposed to cooperation. It also provides an example where inequality and inefficiency are positively correlated.

1 Introduction

The purpose of the paper is to analyse the problem of regional inequality in a country which consists of several unequally developed regions. In particular, we look at the efficiency aspects of regional inequality, how it affects output levels of the regions and how it determines capital accumulation. The paper also looks at the viability of inter-regional cooperation when regional disparities are present. Instances abound all over the globe of regional forces springing up and demanding autonomy. They range from violent separatist uprises like in Northern Ireland, Northern Sri Lanka and in the North Indian State of Kashmir, to relatively peaceful demands for fiscal autonomy by a region within a country. As examples of the latter, we may look at the newly emerging states of Uttaranchal, Jharkhand or Chhattisgarh within the Indian Union. In this paper we shall be concerned with the possibility of the second type of regional non-cooperation. More specifically we try to find out the economic causes as to why a particular region would wish to have fiscal autonomy as opposed to cooperation with the rest of the country. A related question which is also examined in this paper is that whether a transfer of income from an economically advanced to an economically backward region is necessary for the viability of inter-regional cooperation.

We confine our analysis to less developed economies. Consequently, we model our representative economy as one where there is surplus unemployed labour at an institutionally fixed wage rate. The capital stock, on the other hand, is fully employed and acts as the primary constraint to development.

Also, it is in terms of the regional stocks of capital that different regions within the economy are said to be unequally developed. The link between the different regions is established through the production and consumption of a common public good. The public good, by its very nature, is enjoyed by each region, irrespective of its contribution to production. On the other hand, so far as production of the public good is concerned, in a situation of regional fiscal autonomy, each region is allowed to choose how much of its resources should be devoted to the production of the common public good. We have compared the situation of regional fiscal autonomy with another situation where the country is conceived as one single unit and the decision as to how much resources are to be devoted to the production of the public good is centrally taken. The latter situation is referred to as one with inter-regional cooperation.

We show that in the absence of any inter-regional transfer, the economically backward regions have strong reasons to prefer regional fiscal autonomy to inter-regional cooperation, if not immediately, then surely at some point in time in the future. Our analysis also suggests that in order to make inter-regional cooperation viable, transfer of income from the rich to the poorer regions might be necessary. Since inter-regional cooperation yields higher income for the country as a whole, such transfers can be justified on grounds of efficiency.

2 The model

We consider an economy consisting of two regions indexed by $i = 1, 2$. In each region, two goods are produced, a private good denoted by X and a public good denoted by G. Each good is produced with two factors of production, labour(L) and capital(K). In each region, the supply of labour is infinitely elastic at a given wage rate which is assumed to be fixed in terms of the private good. The stocks of capital are given in each region at any point in time and are fully employed. For simplicity, it is assumed that technology is the same across regions. Also, the capital-labour ratios are assumed to be fixed in each sector and assumed to be the same across sectors. By appropriately choosing units, we can take the capital labour ratio in each sector equal to unity. Let one unit of labour and one unit of capital produce one unit of final output in the public good sector; and let a units of labour and a units of capital produce one unit of final output in the private good sector. Then in any given period, for the ith region we have

$$K_i = G_i + aX_i = L_i, i = 1, 2 \tag{1}$$

where L_i, the level of employment in region i, is endogenously determined by the given stock of capital K_i. Let G be the total amount of public good produced in the economy which is the sum of public goods produced in each region, i.e.,

$$G = G_1 + G_2 \tag{2}$$

We assume that the public good is not directly consumed: instead it affects consumption indirectly by increasing productivity in the private good sector. Basically, we are thinking of the public good as an infrastructural facility which improves production technology in the private good sectors of both regions. More specifically we assume that

$$a = G^{-s}, 0 < s < 1 \tag{3}$$

where s is a constant. Equation (3) implies that the output-capital ratio and the output labour ratio rise in the private sector as G rises. Since s is assumed to lie between zero and one, the output-input ratios rise at a diminishing rate with increases in G.

We first consider equilibrium under regional autonomy. This is the situation where each region can independently choose how much of its resources would it devote to the production of the public good. The result is a non-cooperative Nash equilibrium where the ith region maximizes X_i by choosing G_i taking the level of public good production in the other region as given.

Let $K_{ix} = aX_i$ be the amount of capital devoted to the production of the private good in the ith region. Then by definition $X_i = \frac{K_{ix}}{a}$ and using equations (1) and (3), this may be written as

$$X_i = (K_i - G_i)G^s \tag{4}$$

Since the ith region takes the level of public good production of the other region as given while maximizing X_i, we have $\frac{dG}{dG_i} = 1$. Hence from the first order condition of maximization we get

$$G = sK_{ix} \tag{5}$$

It may be easily verified that the second order condition is satisfied for $0 < s < 1$. Equation (5) immediately implies that in equilibrium $K_{1x} = K_{2x}$ so that with regional autonomy we get $X_1 = X_2$, i.e. the levels of private goods production are equalized across regions.

The next question is: how to finance G_i? We assume that G_i is financed by an ad velorem tax t_i on the private good. Hence assuming perfect competition, the zero-profit condition for the private good sector in the ith region can be written as

$$1 = (1 + t_i)(w + r_i)a \tag{6}$$

where the private good is taken to be the numeraire with its price equal to unity and where r_i is the return to a unit of capital in region i and the same wage rate w (fixed in terms of the private good)is assumed to prevail in both regions. Clearly from equation (6), r_i can be determined if a and t_i were known. Also the cost of producing a unit of public good in region i is given by $(w + r_i)$. Therefore, from the balance of the ith regional government's budget we have

$$(w + r_i)G_i = t_i(w + r_i)aX_i$$

which implies that

$$G_i = t_i K_{ix} \tag{7}$$

Equations (7) and (1) together yield

$$K_{ix} = \frac{K_i}{1 + t_i} \tag{8}$$

$$G_i = \frac{t_i}{1 + t_i} K_i \tag{9}$$

Since $G = G_1 + G_2 = (t_1 + t_2)K_{ix} = sK_{ix}$, we have

$$t_1 + t_2 = s \tag{10}$$

Also from equation (8) we get

$$\frac{K_1}{K_2} = \frac{1 + t_1}{1 + t_2} \tag{11}$$

From (10) and (11) t_1 and t_2 can be solved as

$$t_1 = \frac{K_1(1 + s) - K_2}{K}, t_2 = \frac{K_2(1 + s) - K_1}{K} \tag{12}$$

where $K = K_1 + K_2$. Consequently, the solutions for G, K_{ix} and G_i are given by

$$G = \frac{s}{2 + s} K, K_{ix} = \frac{K}{2 + s}, G_i = K_i - \frac{K}{2 + s} \tag{13}$$

Since the existence of an interior solution is guaranteed if and only if $t_i > 0, G_i > 0$, from equation (12) the necessary and sufficient condition for the existence of an interior solution is obtained as

$$1 + s > \frac{K_1}{K_2} > \frac{1}{1 + s} \tag{14}$$

The interpretation of condition (14) is straight forward. Condition (14) implies that the capital stocks of the two regions will have to be sufficiently close in order that an interior solution exists: for if the capital stock of one of the regions is very large relative to the other, then the region with the larger capital stock will devote so much resources to the production of the public good that the smaller region will find it optimal not to produce any public good at all. A priori there is no reason for condition (14) to hold. However, for the time being we shall assume that it holds and later we shall discuss what happens if this condition is not satisfied.

Let us now consider the behaviour of the economy over time. We assume that a fraction b_i of X_i, $b_i = b(X_i), b' > 0$, is saved and invested in each region. Since $X_1 = X_2$, therefore $b_1 = b_2$ and saving and investment, denoted by I, is the same across regions. Let us assume that the capital stock does not depreciate. Also let us assume, without any loss of generality, that $K_1 > K_2$. Then the relative capital stocks in the next period satisfies

$$\frac{K_1 + I}{K_2 + I} > \frac{K_1}{K_2}$$

which implies that disparities in regional capital stocks fall over time. Since regional capital stocks determine regional levels of employment, over time disparities in the levels of employment across regions go down as well. Thus we have the following proposition:

Proposition 3 *In an equilibrium with regional fiscal autonomy, if both regions produce positive amounts of the public good, income and consumption are equalized across regions at each period of time and disparities in capital stocks and employment across regions fall over time.*

Next we consider the situation where the two regions are combined and treated as one single unit. Consequently, the decision as to how much public good is to be produced is centrally taken. The result is a cooperative equilibrium. Let the variables in the cooperative equilibrium be denoted by a star. The problem is then to maximize $X^* = X_1^* + X_2^*$ subject to $K = G^* + K_x^*$ and equation (3). The maximization exercise yields

$$G^* = sK_x^* = s(K_{1x}^* + K_{2x}^*)$$

Since the two regions are now treated as one single unit, a uniform tax t is imposed in both regions. We, therefore, have

$$K_{ix}^* = \frac{K_i}{1+t}, G_i^* = \frac{t}{1+t}K_i \tag{15}$$

Also we get

$$G^* = \frac{t}{1+t}K = s(K_{1x}^* + K_{2x}^*) \tag{16}$$

which implies that $t = s$. Hence the final solutions in the cooperative equilibrium are obtained as

$$K_{ix}^* = \frac{K_i}{1+s}, G_i^* = \frac{s}{1+s}K_i, G_i^* = \frac{s}{1+s}K \tag{17}$$

Comparing the expressions for G^* and G from equations (12) and (17) it immediately follows that $G^* > G$, i.e. more public good is produced under regional cooperation. Also

$$\frac{X_1^*}{X_2^*} = \frac{K_{1x}^*}{K_{2x}^*} = \frac{K_1}{K_2} \tag{18}$$

which implies that under regional cooperation, the region with the higher capital stock will produce more private good. Given our assumption regarding savings and investment, this, in turn, implies that the level of investment will be higher in the region with the higher stock of capital. Hence we have

$$\frac{K_1 + I_1}{K_2 + I_2} < \frac{K_1}{K_2}$$

where I_i is the level of investment in region i and by assumption $K_2 > K_1$. Thus we have the following proposition

Proposition 4 *In an equilibrium with regional cooperation, the level of income (measured in terms of the amount of the private good produced) will be higher in the region with the higher stock of capital. Also regional inequality will increase over time.*

In spite of increasing inequality, a particular region can still gain from cooperation. From equations (12) and (17) it is straight forward to see that

$$X_i = (\frac{s}{2+s})^s K^s \frac{K}{2+s}, X_i^* = (\frac{s}{1+s})^s K^s \frac{K_i}{1+s}$$

so that the ith region (weakly) gains from cooperation if and only if

$$\frac{K_i}{K} \geq (\frac{1+s}{2+s})^{1+s} \tag{19}$$

Of course, this comparison is valid provided interior solutions exist for the non-cooperative equilibrium, i.e. provided

$$\frac{K_i}{K} > \frac{1}{2+s}$$

It is straight forward to show that

$$\frac{1}{2+s} < (\frac{1+s}{2+s})^{1+s}$$

for $0 < s < 1$. Thus a region loses from cooperation if

$$\frac{1}{2+s} < \frac{K_i}{K} < (\frac{1+s}{2+s})^{1+s} \tag{20}$$

It is quite possible that initially condition (19) is satisfied for both regions and each gains from cooperation. However, as we have seen above, in an equilibrium with regional cooperation the relative capital stock of the initially backward region falls over time. So a time might come when it falls sufficiently such that condition (20) is satisfied. In other words, the backward region might not find the non-cooperative equilibrium immediately attractive. However, at some point in time in the future the non-cooperative equilibrium might be found preferable to the cooperative equilibrium by the backward region. This perhaps partly explains why in a particular country a backward region might stick to cooperation for a long time and then start demanding autonomy.

Even though a particular region might lose from cooperation, the country as a whole must gain from it. In other words, it is always true that $X_1 + X_2 < X_1^* + X_2^*$ which immediately implies that the relatively advanced region always gains from cooperation. Another implication of this is that it is always possible to divide the gains from cooperation between the regions such that no region is worse off compared to the non-cooperative equilibrium. This might entail transfer of income from the advanced to the backward region. Indeed, in the absence of such transfers it might be difficult to pursue the backward region to stick to cooperation. Since cooperation is income increasing for the country as a whole and is definitely in the interest of the advanced region, such transfers can be justified on grounds of efficiency. Altetrnatively we might say that our model provides an example where inequality is inefficient and to remove this inefficiency inter-regional transfers are necessary.

A couple of comments on the above model are now in order. First consider the case where condition (14) is not satisfied and we have $\frac{K_1}{K} \leq \frac{1}{2+s}$. This is a situation where region 1 is too small to produce any public good in the non-cooperative equilibrium. It can be easily verified that in this case $K_{1x} = K_1$ and $G = \frac{s}{1+s}K_2$. The cooperative solution, of course, remains the same as before. Therefore

$$\frac{X_1}{X_1^*} = (1+s)(\frac{K_2}{K})^s \tag{21}$$

Since $\frac{K_1}{K} \leq \frac{1}{2+s}$, it follows that $\frac{K_2}{K} \geq \frac{1+s}{2+s}$ so that from the last equation we get

$$\frac{X_1}{X_1^*} \geq (1+s)(\frac{1+s}{2+s})^s \tag{22}$$

It is straight forward to show that the right hand side of the above equation is greater than unity provided $0 < s < 1$. Hence in this case the backward region always gains by switching to the non-cooperative equilibrium. By a similar analysis it can also be shown that the advanced region always gains by cooperation. Thus the conclusions that we derived for the case with interior solutions of the non-cooperative equilibrium hold with greater strength if no interior solution exists. The only difference is that in the latter case even in the non-cooperative equilibrium capital stocks of the two regions diverge over time. This gives an additional reason why the backward region would like to switch to the non-cooperative equilibrium as soon as possible; for if the switching is delayed and its relative capital stock falls by a sufficiently large amount, it can never get level with the advanced region.

Secondly, one can easily find out the implications of inter-regional capital mobility in the above model. Suppose capital moves to that region where it earns a higher return. Our previous analysis suggests that in a non-cooperative equilibrium the backward region has a lower tax rate (see equation (11) above) and hence a higher rate of return to capital, since wages are the same across regions (follows from equation (6)). This implies that in the non-cooperative equilibrium, capital will move from the advanced region to the backward region so that regional disparities will tend to vanish more quickly. On the other hand, in the cooperative equilibrium, since the same tax is imposed in both regions, capital will have no incentive to move. So regions will diverge as indicated above. Thus capital mobility strengthens the conclusions we have obtained before.

3 Concluding remarks

In this paper we have analysed the viability of inter-regional cooperation in a country which consists of unequally developed regions. It is shown that a backward region will have strong reasons for not preferring regional cooperation. It is also shown that in order to make inter-regional cooperation viable, transfer of income from the advanced to the backward region might be necessary.

References

1. Friedman JW (1977) Oligopoly and the Theory of Games, North-Holland
2. Krugman P (1981) Trade, Accumulation and Uneven Development, Journal of Development Economics, 8(2):149-161

Part III

Historical Notes

A Brief History of Economics: An Outsider's Account

Bikas K Chakrabarti

Theoretical Condensed Matter Physics Division and Centre for Applied Mathematics and Computational Science, Saha Institute of Nuclear Physics, 1/AF, Bidhannagar, Kolkata 700 064, India. bikask.chakrabarti@saha.ac.in

When physics started to develop, say with Galileo Galelei (1564-1642), there were hardly any science at a grown-up stage to get help or inspiration from. The only science that was somewhat grown up was mathematics, which is an analytical science (based on logic) and not synthetic (based on observations/ experiments carried out in controlled environments or laboratories). Yet, developments in mathematics, astronomical studies in particular, had a deep impact in the development of physics, of which the (classical) foundation was almost single-handedly laid down by Isaac Newton (1643-1727) in the seventeenth and early eighteenth century. Mathematics remained at the core of physics since then. The rest of "main stream" sciences, like chemistry, biology etc all tried to get inspiration from, utilize, and compare with physics since then.

To my mind, development in social sciences started much later. Even the earliest attempt to model an agricultural economy in a kingdom, the "physiocrats' model", named after the profession of its pioneer, the french royal physician Francois Quesnay (1694-1774), came in the third quarter of the eighteenth century when physics was already put on firm ground by Newton. The physiocrats made the observation that an economy consists of the components like land and farmers, which are obvious. Additionally, they identified the other components as investment (in the form of seeds from previous savings) and protection (during harvest and collection, by the landlord or the king). The impact of the physical sciences, in emphasizing these observations regarding components of an economy, is clear. The analogy with human physiology then suggested that, like the healthy function of a body requiring proper functioning of each of its components or organs and the (blood) flow among them remaining uninterrrupted, each component of the economy should be given proper care (suggesting rent for land and tax for protection!). Although the physiocrats' observations were appreciated later, the attempt to conclude using the analogy with human physiology was not.

Soon, at their last phase, Mercantilists, like Wilhelm von Hornick (1638-1712), James Stewart (1712-1780) et al, made some of the most profound and emphatic observations in economics, leading to the foundation of political economy. In particular, the observations by the British merchants (who traded in the colonies, including India, in their own set terms) that instability/unemployment growing at their home country in years whenever there had been a net trade deficit and out-flow of gold (export being less than import). This led to the formulation of the problem of effective demand: even though the merchants, or traders were independently trading (exporting or importing goods) with success, the country's economy as a whole did not do well due to lack of overall demand when there was a net flow of gold (the international exchange medium) to balance the trade deficit! This remains still a major problem in macroeconomics. The only solution in those days was to introduce tax on import: the third party (namely the government) intervention on individuals' choice of economic activity (trade). This immediately justified the involvement of the government in the economic activities of the individuals.

In a somewhat isolated but powerful observation, Thomas Malthus (1766-1834) made a very precise modelling of the conflict between agricultural production and population growth. He assumed that the agricultural production can only grow (linearly) with the area of the cultivated land. With time t, say year, the area can only grow linearly ($\propto t$) or in arithmetic progression (AP). The consumption depends on the population which, on the other hand, grows exponentially ($\exp[t]$) or in geometric progression (GP). Hence, with time, or year $1, 2, 3, \ldots$, the agricultural production grows as $1, 2, 3, \ldots$, while the consumption demand or population grows in a series like $2, 4, 8, \ldots$. No matter, how much large area of cultivable land we start with, the population GP series soon takes over the food production AP series and the population faces a disaster — to be settled with famine, war or revolution! They are inevitable, as an exponentially growing function will always win over a lineraly growing function and such disasters will appear almost periodically in time!

Adam Smith (1723-1790) made the first attempt to formulate the economic science. He painstakingly argued that a truely many-body system of selfish agents, each having no idea of benevolence or charity towards its fellow neighbours, or having no foresight (views very local in space and time), can indeed reach an equilibrium where the economy as a whole is most efficient; leading to the best acceptable price for each commodity. This 'invisible hand' mechanism of the market to evolve towards the 'most efficient' (beneficial to *all* participating agents) predates by ages the demonstration of 'self-organisation' mechanism in physics or chemistry of many-body systems, where each constitutent cell or automata follows very local (in space and time) dynamical rules and yet the collective system evolves towards a globally 'organised' pattern (cf. Ilya Prigogine (1917-), Per Bak (1947-2002) et al). This idea of 'self-organizing or self-correcting economy' by Smith of course contradicted the prescription of the Mercantilists regarding government intervention in the economic activities of the individuals, and argued tampering by any external agency to be counterproductive.

Soon, the problem of price or value of any commodity in the market became a central problem. Following David Ricardo's (1772-1823) formulation of rent and labour theory of value, where the price depends only on the amount of labour put by the farmers or labourers, Karl Marx (1818-1883) formulated and forwarded emphatically the surplus labour theory of value or wealth in any economy. However, none of them could solve the price paradox: why diamond is costly, while coal is cheap? The amount of labour in mining etc are more or less the same for both. Yet, the prices are different by astronomical factors! This clearly demonstrates the failure of the labour theory of value. The alternative forwarded was the utility theory of price: the more the utility of a commodity, the more will be its price. But then, how come a bottle of water costs less than a bottle of wine? Water is life and certainly has more utility! The solution identified was marginal utility. According to marginal utility theory, not the utility but rather its derivative with respect to the quantity determines the price: water is cheaper as its marginal utility at the present level of its availability is less than that for wine — will surely change in a desert. This still does not solve the problem completely. Of course increasing marginal utility creates increasing demand for it, but its price must depend on its supply (and will be determined by equating the demand with the supply)! If the offered (hypothetical) price p of a commodity increases, the supply will increase and the demand for that commodity will decrease. The price, for which supply S will be equal to demand D, will be the market price of the commodity: $S(p) = D(p)$ at the market (clearing) price. However, there are problems still. Which demand should be equated to which supply? It is not uncommon to see often (in India) that price as well as the demand for rice (say) increases simultaneously. This can occur when the price of the other staple alternative (wheat) increases even more.

The solutions to these problems led ultimately to the formal development of economic science in the early twentieth century by Léon Walras (1834-1910), Alfred Marshal (1842-1924) and others: marginal utility theory of price and cooperative or coupled (in all commodities) demand and supply equations. These formulations went back to the self-organising picture of any market, as suggested by Adam Smith, and incorporated this marginal utility concept, and utilized these coupled demand-supply equations: $D_i(p_1, p_2, \ldots, p_i, \ldots, p_N, M) = S_i(p_1, p_2, \ldots, p_i, \ldots, p_N, M)$ for N commodities and total money M in the market, each having relative price tags p_i (determined by marginal utility rankings) and demand D_i and supply S_i; $i = 1, 2, \ldots, N$ and the functions D or S are in general nonlinear in their arguments. These formal and abstract formulations of economic science were not appreciated very much in its early days and had a temporary setback. The lack of acceptance was due to the fact that neither utility nor marginal utility is measurable and the formal solutions of these coupled nonlinear equations in many (p_i) variables still remain elusive. The major reason for the lack of appreciation for these formal theories was a profound and intuitive obsevation by John Maynard Keynes (1883-1946) on the fall of aggregate (or macroeconomic) effective demand in the market (as pointed out earlier by the Mercantilists; this time due to 'liquidity preference' of money by the market participants) during the great depression of 1930's. His prescription was for government intervention (in direct contradiction with the 'laissez-faire' ideas of leaving the market to its own forces to bring back the equilibrium, as Smith, Walras et al proposed) to boost aggregate demand by fiscal measures. This prescription made immediate success in most cases. By the third quarter of the twentieth century, however, its failures bacame apparent and the formal developments in microeconomics took the front seat again.

Several important, but isolated observations in the meantime contributed later very significantly. Vilfredo Pareto (1848-1923) observed that the number density $P(m)$ of riches in any society decreases rather slowly with their richness m (measured in those days by palace sizes, number of horses, etc of the kings/landlords in all over Europe): $P(m) \sim m^{-\alpha}$; for very large m (very rich people); $2 < \alpha < 3$ (Cours d'Economic Politique, Lausanne, 1897). It may be mentioned, at almost the same time, Joshiah Willard Gibbs (1839-1903) had put forward precisely that the number density $P(\epsilon)$ of particles (or microstates) with energy ϵ in a thermodynamic ensemble in equilibrium at temperature T falls off much faster: $P(\epsilon) \sim \exp[-\epsilon/T]$ (Elementary Principles of Statistical Mechanics, 1902). This was by then rigorously established in physics. The other important observation was by Louis Bachelier (1870-1946) who modelled the speculative price fluctuations (σ), over time τ, using a Gaussian statistics (for random walk): $P(\sigma) \sim \exp[-\sigma^2/\tau]$ (Thesis: Théorie de la Spéculation, Paris, 1900). This actually predated Albert Einstein's (1879-1955) random walk theory (1905) by five years. In another isolated development, mathematician John von Neumann (1903-1957) started developing the game theories for microeconomic behavior of partners in oligopolistic competitions (to take care of the strategy changes by agents, based on earlier performance).

In the mainstrem economics, Paul Samuelson (1915-) investigated the dynamic stabilities of demand-supply equilibrium by formulating, following Newton's equations of motion in mechanics, dynamical equations $\frac{dD_i}{dt} = \sum_i J_{ij} D_j(p_1, p_2, \ldots, p_N, M)$ and $\frac{dS_i}{dt} = \sum_i K_{ij} S_i(p_1, p_2, \ldots, p_N, M)$, with the demand and supply (overlap) matrices $\underline{J}$ and $\underline{K}$ respectively for N commodities, and by looking for the equilibrium state(s) where $dS/dt = 0 = dD/dt$ at the market clearing prices $\{p\}$ where $D_i(\{p\}, M) = S_i(\{p\}, M)$. Jan Tinbergen (1903-1994), a statistical physicist (student of Paul Ehrenfest of Leiden University) analysed the business cycle statistics and initiated the formulation of econometrics. By this time, these formal developments in economics, with clear impact of other developed sciences (physics in particular), were getting recognized. In fact, Tinbergen was the first recipient of the newly instituted Nobel prize in Economics in 1969 (for other sciences, they started in 1901; a delay by 68 years in 105 years' history of the prize!) and the next year, the prize went to Samuelson. Soon, the formal developments like the axiomatic foundations of utility (ranking) theory, and solution of general equilibrium theory by Kenneth Arrow (1921-), those of George Stigler (1911-1991), who first performed Monte Carlo simulations of markets (similar to those of thermodynamic systems in physics), or that of John Nash (1928-), giving the proof of the existence of equilibrium solutions in strategic games, etc, all were appreciated by awarding the Nobel prizes in economics (in 1972, 1982 and 1994 respectively). Although the impact of developments in physics had a clear mark in those of economics so far, it was not that explicit until about a decade and a half back.

The latest developments (leading to econophysics) had of course its seed in several earlier observations. Important among them was by Benoit Mandelbrot (1924-) when he observed in 1963 that the speculative fluctuations (in the cotton market for example) have a much slower rate of decay, compared to that suggested by the Gaussian statistics of Bachelier, and falls down following a power law statistics: $P(\sigma) \sim \sigma^{-\alpha}$ with some robust exponent value (α) depending on the time scale of observations. With the enormous amount of stock market data now available on the internet, Eugene Stanley, Rosario Mantegna and coworkers established firmly the above mentioned (power-law) form of the stock price fluctuation statistics in late 1990's. Simultaneously, two important modelling efforts, inspired directly from physics, started: the minority game models, for taking care of contigious behavior (in contrast to perfect rational behavior) of agents in the market, and learning from the past performance of the strategies, were developed by Brian Arthur, Damien Challet, Yi-Cheng Zhang et al, starting 1994. The other modelling effort was to capture the income or wealth distribution in society, similar to energy distributions in (ideal) gases. These models intend to capture both the initial Gamma/log-normal distribution for the income distributions of poor and middle-income groups and also the Pareto tail of the distribution for the riches. It turned out, as shown by the Kolkata group during the last half of 1990 to the first half of 2000, a random saving gas model can easily capture these features of the distribution function. However, the model had several well documented previous, somewhat incomplete, versions available for a long time. Meghnad Saha (1893-1956), the founder of Saha Institute of Nuclear Physics, Kolkata (named so after its founder's death), and collaborators, already discussed at length in their text book, in the 1950's, the possibility of using Maxwell-Boltzmann velocity distribution (a Gamma distribution) in an ideal gas to represent the income distribution in societies: "suppose in a country, the assessing department is required to find out the average income per head of the population. They will proceed somewhat in the similar way ... (the income distribution) curve will have this shape because the number of absolute beggers is very small, and the number of millionaires is also small, while the majority of the population have average income." (section on 'Distribution of velocities' in A Treatise on Heat, M. N. Saha and B. N. Srivastava, Indian Press, Allahabad, 1950; pp. 132-134). This modelling had the obvious drawback that the distribution could not capture the Pareto tail. However, the accuracy of this Gibbs distribution for fitting the income data available now in the internet has been pointed out recently by Victor Yakovenko and collaborators in a series of papers since 2000. The 'savings' ingredient in the ideal-gas model, required for getting the Gamma function form of the otherwise ideal gas (Gibbs) distribution, was also discovered more than a decade earlier by John Angle. He employed a different driver in his stochastic model of inequality process. This inequality coming mainly from the stochasticity, together with the equivalent of saving introduced in the model. A proper Pareto tail of the Gamma distribution comes naturally in this class of models when the saving propensity of the agents are distributed, as noted first by the Kolkata group and analyzed by them and by the Dublin group led by Peter Richmond.

Apart from the intensive involvements of physicists together with a few economists in this new phase of development, a happy feature has been that econophysics has almost established itself as a (popular) research discipline in statistical physics. Many physics journals have started publishing papers in such an interdisciplinary field. Also, courses in econophysics are being offerred in several universities, mostly in their physics departments.

In spite of all these, it must be stated that there has, so far, been no spectacular success. Indeed, the mainstream economists are yet to take note of these developments. In her account, reporting on the Econophys-Kolkata I (New Scientist, UK, 12 March 2005 issue, pp.6-7), Jenny Hogan reported several criticisms by economists, mostly appreciating the observations, but not the modelling efforts! The same kind of criticism have recently been expressed more emphatically by economists Mauro Gallegati, Thomas Lux and others (http://www.unifr.ch/econophysics; doc/0601001; Physica A in press). We have included a few responses by physicists like Peter Richmond and social statistician like John Angle in this volume. Economist J Barkley Rosser, in his intriguing reflections (in the following paper) on these new developments of econophysics, detects the same stories of the past becoming present, predicting the future of econophysics to become a past story of economics soon!

Acknowledgments

I am grateful to Anindya-Sundar Chakrabarti and Arnab Chatterjee for careful reading of this manuscript and useful criticisms.

The Nature and Future of Econophysics

J. Barkley Rosser, Jr.

Department of Economics, James Madison University, Harrisonburg, VA, USA.
rosserjb@jmu.edu

1 Definitional issues

In discussing the nature of econophysics, a primary issue must be to understand what it is. This is a rather complicated matter, but attempts at definition have been made. As the neologizers of the term,[1] Rosario Mantegna and H. Eugene Stanley have a distinct authority in this matter. They have proposed the following to define "the multidisciplinary field of econophysics ...[as] a neologism that denotes the activities of physicists who are working on economics problems to test a variety of new conceptual approaches deriving from the physical sciences" ([2] pp. viii-ix).

What is most striking about this definition is that it is not an intellectual one primarily, but rather a sociological one. It is based on who is doing the working on economics problems, physicists, not specifically on what those problems are or what the specific methods or theories are that they are using from physics to study or solve these problems. The more usual way to define a "multidisciplinary discipline" (or interdisciplinary or transdisciplinary discipline)[2] is to do so in terms of the ideas that it deals with. Arguably this is the case with such well-established entities as political economy or biophysics. Indeed, this more sociologically oriented definition resembles more the contrast between socio-economics and social economics, which seem to deal with nearly

[1] See [[1] p. 225] for an account of its neologization, with Stanley being given specific credit for this innovation at the second Statphys-Kolkata conference in 1995.

[2] It can be argued that "transdisciplinary" might be a better label for econophysics. "Multidisciplinary" suggests distinct disciplines discussing as with an economist and a physicist talking to each other. "Interdisciplinary" suggests a narrow specialty created out of elements of each separate discipline, such as a "water economist" who knows some hydrology and some economics. However, "transdisciplinary" suggests a deeper synthesis of approaches and ideas from the disciplines involved, and is the term favored by the ecological economics for what they are trying to develop.

identical ideas and subject matter, but which differ largely in that the former is largely done by sociologists while the latter is largely done by economists[3].

2 Relations between the disciplines

Given that econophysics involves in effect physicists doing economics with theories from physics, this raises the question of how the two disciplines relate to each other both ideationally and sociologically. In the realm of ideas it has been traditional to argue that they belong to the distinct categories of physical (or natural or "hard") science and social (or "soft") science. However, this partly elides the question as there are areas of physics that are not so "hard" in that they are not able to be tested experimentally (string theory being a current example), while increasingly one finds the use of experiments in economics, even if these are arguably not as controlled or scientific as most experiments in physics, chemistry, or biology.

Another way to view how they relate is in terms of what parts of the universe they study within the great hierarchy of the structure of reality. As one moves from a micro scale perspective to a macro scale perspective we find physics dealing with the two greatest extremes: nuclear physics at the sub-atomic particle level, and astrophysics at the cosmic and universal level, with string theory arguably attempting to deal with both simultaneously. As one moves up the hierarchy from nuclear physics through greater levels of aggregation one moves through realms studied respectively by chemistry, molecular biology, organic biology, psychology, economics, political science and sociology, ecology, climatology and geology, and finally to astrophysics as one moves beyond the scale of the planet earth. In this hierarchy we find the social sciences occupying an intermediate scale, with economics providing a crucial link from the behavior of individual people to the behavior of groups of people.

There is also an intellectual hierarchy of the disciplines as well. This generally moves from the more abstract and theoretical to the less so, with pure mathematics at the top, followed by its applied forms (including statistics and computer science), with physics generally coming next, again with the theoretical viewed as above the empirical, then to chemistry (and closely related molecular biology), then on to economics, which has the most mathematical orientation of the social sciences, on through ecology, psychology, and to political science and sociology, even as these latter disciplines have become more mathematical in some of their branches, especially with the application of game theory.

[3] This is symbolized by two competing professional associations. The Society for the Advancement of Socio-Economics (SASE) is dominated by sociologists, although there are economists associated with it, while the Association for Social Economics (ASE) is essentially the reverse. The ASE holds meetings in conjunction with the American Economic Association annual meetings, while SASE does not.

This implicit intellectual superiority of physics over economics opens the door to potential conflicts between the two groups as the project of econophysics has developed and proceeded[4]. Thus we can record prejudices that the two groups entertain regarding each other. On the one hand, econophysicists argue that regular economists are unwilling to accept or deal with data or facts that do not conform to the predictions of their outmoded theories, that in this regard economists are not really scientists. On the other hand, economists who study the same data and models that the econophysicists do have been known to complain that the econophysicists are not always aware of the work that the economists have done or the true nature of their theories and that the econophysicists apply models to data without any real theory at all. Arguably, the future development of econophysics will involve the overcoming of these prejudices through communication and mutual research and effort.

3 Problems being studied, methods being used

Given that econophysicists work on economic problems with new conceptual approaches from the physical sciences, the issue arises as to what these are. Perhaps the most intensively studied have been the distributions of returns in financial markets [3–8]. Also studied have been the distribution of income and wealth [9–11], the distribution of economic shocks and growth rate variations [12,13], the distribution of firm sizes and growth rates [14,15], the distribution of city sizes [16], and the distribution of scientific discoveries [17]. Leading models have come from statistical mechanics [18], earthquake models [8], and self-organized criticality models of sandpile avalanche dynamics [19].

Arguably the clearest contrast between the approach of econophysicists and more regular economists has been in the conviction of the former that many of these phenomena can be better described using scaling laws that imply non-Gaussian distributions exhibiting skewness and leptokurtosis rather than Gaussian distributions. A major area of contention regarding this has been in the area of financial returns distributions. The Gaussian theory dates to Bachelier in 1900 [20], who developed the theory of Brownian motion five years before Einstein did in order to model financial markets. This approach would develop through the mean-variance approach to risk analysis and culminate in such conceptual outcomes as the Black-Scholes formula [21], widely used for the pricing of options and derivatives in financial markets.

The contrasting approach using scaling laws derives initially from Pareto in 1897 [22], who used it to study income distribution, and was first applied to financial markets by Mandelbrot in 1963 [23]. The Paretian distribution was also used by Lotka [24] to study the pattern of scientific discoveries and

[4] Recognition of this by economists has led Mirowski [3] to accuse economists of suffering from "physics envy"

by Zipf [25] to study city size distributions, as well as by Ijiri and Simon [26] to study firm size distributions.

4 But who preceded whom?

However, at this point we come to a deeper issue and problem: it is not clear who preceded whom disciplinarily in these developments. Indeed, what we have is a truly complicated muddle of developments. What seem to be models of physics in some cases came from economics, and the much-derided standard models of economics largely came from physics. "Econophysics" may well simply be a new name for something that has been going on for a long time.

Thus, Vilfredo Pareto was an economist and sociologist, even as he preceded the physicists in studying distributions exhibiting scaling laws. Bachelier was a mathematician who studied financial markets, but developed the idea of Brownian motion before the physicists got to it. On the other hand, it was a physicist, Osborne [27], who played a major role in enforcing that standard financial economics would use the Gaussian Brownian motion model.

The role of physicists and physics models as the foundations for the standard neoclassical model that the current econophysicists seek to displace is a much studied story [3]. Canard in 1801 [28] first proposed that supply and demand were ontologically like contradicting physical forces. The founder of general equilibrium theory in economics, Léon Walras [29] was deeply influenced by the physicist Louis Poinsot [30] in his formulation of this central concept[5]. The father of American mathematical economics in its neoclassical form, Irving Fisher [32], was a student of the father of statistical mechanics, J. Willard Gibbs [33]. The culmination of this transfer of essentially nineteenth century physical concepts into standard neoclassical economics came with the publication of Paul Samuelson's *Foundations of Economic Analysis* in 1947 [34], Samuelson himself having originally been an undergraduate physics major at the University of Chicago. Furthermore, even the later, more direct introduction of statistical mechanics applications into economics originally came from economists themselves [35].

5 Is econophysics heterodox?

Clearly econophysics is an ongoing enterprise; whatever it is and whoever is doing it is not going to stop in the near future, whatever its longer term evolution. This raises a more particular question regarding its relationship with the economics profession. In particular, is it something that is acceptable to the mainstream or to orthodox economic theory, or is it something that is

[5] See [31] for further discussion of this point.

outside of this mainstream, something that is heterodox. And if so, will this remain the case?

Such a question requires some defining of terms. I shall follow the categorization and set of definitions provided in [36]. These authors argue that there are two aspects involved in understanding useful definitions of these terms, intellectual and sociological, the same two that have been hovering over our discussion already. Thus, they argue that orthodox refers purely to an intellectual category. Orthodox economic theory is an established and more or less internally consistent set of ideas that have been widely accepted, at least in the recent past. The clearest way to identify this set of ideas is to see what is presented in the most widely used textbooks, with perhaps its purest expression being found in undergraduate textbooks rather than graduate ones, which may be more likely to contain newer cutting edge ideas that may fit less well with this established orthodoxy. Indeed, an important aspect of this orthodoxy is that it is not generally cutting edge or new. It may well be something that is already partially fossilized, something to teach undergraduate students, but not something to be taken too seriously by the most advanced researchers at the cutting edge of the profession.

At this point it should probably be noted that when econophysics sometimes criticize what they consider to be economic theory, it is sometimes this established orthodoxy that is widely publicized, but which is not taken entirely seriously by many advanced economic researchers. Many of them are fully aware of its limitations and problems. This is part of the frustration that some economists have in communicating with some econophysicists.

It should also be noted that even though the established orthodoxy may have a degree of internal consistency, this is likely to be ultimately incomplete. As is well known, some of the most important advances in all sciences come from trying to resolve unresolved contradictions within normal science theories [37]. The degree to which orthodox economic theory is internally consistent is generally exaggerated when it is taught to students, especially undergraduate students, perhaps to make it simpler for them, if not necessarily for their teachers, although it is undoubtedly the case that some of the teachers believe fully in the consistency or truth of what they are teaching. Nevertheless, it is usually after an orthodoxy (such as "neoclassical economics") has already come to be formed or established that some critic puts a label on it and it comes to appear to be the unified whole that it really is not [38].

"Mainstream" refers to the economics done by those economists who are the dominating figures in the profession, hold professorships at the most prestigious universities and institutes, who edit the most highly ranked journals, who control funding of research at the highest levels, win Nobel Prizes, and so forth. Many critics who are not very well informed believe that this set of people also believe in or strongly advocate the ideas of the most well-known established orthodoxy. Surprisingly this is not always the case, indeed is more likely not to be the case, although they may defend orthodoxy against what they consider to be poorly thought-out or inappropriate attacks or criticisms.

There is likely to be a show me what is better attitude among the most seriously scientific members of this group, with much less adherence to any particular set of views or theories than many observers think is the case, much more open-mindedness.

Kenneth Arrow is a good example of such a figure, a senior and prominent Nobel Prize winner in economics, he was jointly responsible with Grard Debreu for one of the centerpieces of established neoclassical orthodoxy, the formal proof of the possible existence of a competitive general economic equilibrium [39]. Nevertheless, in the late 1980s, he was responsible along with the physicists such as Philip Anderson in establishing the Santa Fe Institute, which has been a center of the study of such decidedly non-orthodox ideas as nonlinear complex dynamics and a main center for work on econophysics as well, all of which he has been quite supportive of. Kenneth Arrow is as mainstream as one can get sociologically, but he is no simple defender of established orthodoxy in terms of ideas or research in economics. Other figures who are arguably mainstream but not orthodox are interviewed and their work discussed in [36].

Which brings us to "heterodox". In contrast to the previous two, this is seen as being both an intellectual and a sociological category. It is intellectual in being anti-orthodox, or at least non-orthodox. It is also sociologically non-mainstream, its practitioners are outside the dominating group that controls the profession. There is a sense of alienation, even at times persecution, felt within this group. They may feel that their exclusion is due to their papers not being accepted in leading journals due to narrow minded discrimination and unfair exclusion, with the result being their inability to obtain tenure or promotion at good universities or institutes. To the extent that this is tied to ideological factors, such as advocacy of Marxism during various periods, this can add an extra politically vehement element to the sense of alienation and controversy.

Now it must be admitted that there has been such discrimination and rejection of work by some economists who have belonged to certain heterodox schools of economic thought, with Marxists being the most obvious example, and some of these cases are examined and considered in [36]. Efforts at certain universities to purge in a wholesale manner of shut down departments that are controlled by identifiably heterodox economics continue to go on. Curiously, it has been observed that those involved in these kinds of persecutions and discrimination are frequently second or third tier members of the profession, not its leading lights or most advanced researchers. It is the mediocre who tend to be the enforcers of stale orthodoxy[6].

[6] In [36], Kenneth Arrow reports that when the then heterodox Marxist economist, Samuel Bowles, was denied tenure at Harvard, the three members of the department who had served as presidents of the American Economic Association (including Arrow) all supported Bowles receiving tenure.

It can be noted that on the intellectual side there are various schools of economic heterodoxy that have become established to varying degrees over time, such as Marxism, Austrianism, institutionalism, and others. Many of these schools have themselves created their own sub-orthodoxies and mini-mainstreams, with certain universities, institutes, journals, and even funding sources that favor and support them. Ironically, many of these can be as narrow-minded, or even worse, than the established orthodoxy and mainstream in terms of resisting ideas or researchers who challenge their particular shibboleths.

So, now we can approach answering the question of this section: is econophysics heterodox? At the moment the answer is probably "yes". Certainly it challenges the old established orthodoxy of unique equilibria, rational agents, and normality that one finds in the more simple-minded economics textbooks[7]. It is also the case that as most of its practitioners are almost by definition, and certainly in practice, actually professional physicists, they are outside the dominating mainstream group of economists. However, the example of Kenneth Arrow and the important role of the Santa Fe Institute, not to mention the past history of physics ideas becoming incorporated deeply into the fabric of the most established of economic orthodoxy suggest, econophysics has a good chance of ceasing to be heterodox, and quite possibly in the reasonably near future. Many serious economists take the research seriously already. More are likely to do so in the future as more economists become more aware of it, especially if it is successful in explaining phenomena that established economic theories have not done well at explaining. It may be quite sometime, if ever, that the ideas of econophysics become fully orthodox. But as more well known and influential economists become involved in joint research with practicing econophysics and it comes to be published in more established journals in economics, it could well enter into the mainstream in the relatively near future, may in fact already be beginning to do so, if still only the "non-orthodox mainstream".

6 The past as future

This observer is not going to forecast that there will be ideas developed by economists that will influence physics in the future, although this cannot be ruled out. However, the complicated nature of the interplay between economics and physics that has transpired in the past seems very likely to be a model for the future. This complicated interplay replicates a broader pattern of developments as various disciplines are influencing each other through complexity theory and other multi or transdisciplinary ideas. The interactions between physicists and economists that have happened at the Santa Fe Institute echo

[7] An example of an econophysicist very sharply critiquing established economic orthodoxy is [40].

earlier such interactions at institutes in Brussels and Stuttgart. Furthermore, the interactions are not only between economists and physicists but between biologists and both of them as well as some other disciplines.

While physics has probably had the more dominating effect on the development of formal economic theory, a subtext throughout the last two centuries has been a competing strain of interaction between economics and biology, especially evolutionary theory [41]. Alfred Marshall, the main expositor of neoclassical economics in the English language tradition, famously argued ([42], Preface) that biology is the Mecca of economics, even as he codified the earlier mathematical developments drawn from physics in his formal apparatus. This reflected the great influence on Marshall of Darwin. However, earlier Darwin himself had been crucially influenced by the economist Malthus, whose theory of population was a crucial inspiration for Darwin in developing his theory of natural selection. Even so, the lack of a formal mathematical theory of evolution during this earlier period pushed the more biological approach aside, leaving it to the institutionalists [43] and others to apply it in economics until more recently.

However, at the Santa Fe Institute and other places, modern evolutionists have become students of nonlinear complex dynamics and game theory and other more mathematically sophisticated approaches [44], partly as a result of interacting with both physicists and economics. Robert May, a biologist with a physics background played a crucial role in suggesting the application of chaos theory to economics [45][8]. The appearance of ecological economics and its enormous increase in influence in a relatively short time is an indicator of what is going on [47].

References

1. Chakrabarti BK (2005) Econophys-Kolkata: a short story, in Econophysics of Wealth Distributions, Eds. Chatterjee A, Yarlagadda S, Chakrabarti BK, Springer, Milan, pp. 225-228.
2. Mantegna RN, Stanley HE (2000) An Introduction to Econophysics: Correlations and Complexity in Finance. Cambridge University Press, Cambridge.
3. Mirowski P (1989) More Heat than Light: Economics as Social Physics. Cambridge University Press, Cambridge.
4. Mantegna RN (1991) Lévy walks and enhanced diffusion in Milan stock exchange. Physica A 179:232-242.
5. Bouchaud J-P, Cont R (2002) A Langevin approach to stock market fluctuations and crashes. European Physical Journal B 6:543-550.

[8] The neologizing of "econophysics" has since inspired the neologizing of both "econochemistry" and "econobiology", although clearly the latter has been going on already for a long time in practice. In a lecture at Urbino in 2002, Rosser identified a paper by Hartmann and Rössler [46] as a possible example of econochemistry. Of course, Rssler is a physical chemist, much like the late Ilya Prigogine of Brussels.

6. Gopakrishnan P, Plerou V, Amaral LAN, Meyer M, Stanley HE (1999) Scaling of the distributions of fluctuations of financial market indices. Physical Review E 60:5305-5316.
7. Farmer JD, Joshi S (2002) The price dynamics of common trading strategies. Journal of Economic Behavior and Organization 49:149-171.
8. Sornette D (2003) Why Stock Markets Crash: Critical Events in Complex Financial Systems. Princeton University Press, Princeton.
9. Levy M, Solomon S (1997) New evidence for the power-law distribution of wealth. Physica A 242:90-94.
10. Drăgulescu AA, Yakovenko VM (2001) Exponential and power-law probability distributions of wealth and income in the United Kingdom and the United States. Physica A 299:213-221.
11. Chatterjee A, Yarlagadda S, Chakrabarti BK, Eds. (2005) Econophysics of Wealth Distributions, Springer, Milan.
12. Bak P, Chen K, Scheinkman J, Woodford M (1993) Aggregate fluctuations from independent sectoral shocks: self-organized criticality in a model of production and inventory dynamics. Ricerche Economiche 47:3-30.
13. Canning D, Amaral LAN, Lee Y, Meyer M, Stanley HE (1998) A power law for scaling the volatility of GDP growth rates with country size. Economics Letters 60:335-341.
14. Stanley MHR, Amaral LAN, Buldyrev SV, Havlin S, Leschhorn H, Maass P, Salinger MA, Stanley HE (1996) Scaling behavior in the growth of companies. Nature 379:804-806.
15. Takayasu H, Okuyama K (1998) Country dependence on company size, distributions and a numerical model based on competition and cooperation. Fractals 6:67-79.
16. Gabaix, X (1999) Zipf's law for cities: an explanation. Quarterly Journal of Economics 114:739-767.
17. Plerou V, Amaral LAN, Gopakrishnan P, Meyer M, Stanley HE (1999) Similarities between the growth dynamics of university research and competitive economic activities. Nature 400:433-437.
18. Spitzer F (1971) Random Fields and Interacting Particle Systems. American Mathematical Society, Providence.
19. Bak P (1996) How Nature Works: The Science of Self-Organized Criticality. Copernicus Press for Springer-Verlag, New York.
20. Bachelier L (1900) Théorie de la Spéculation. Annales Scientifique de l'École Supérieure III-1:21-86.
21. Black F, Scholes M (1973) The pricing of options and corporate liabilities. Journal of Political Economy 81:637-654.
22. Pareto V (1897) Cours d'Économie Politique. F. Rouge, Lausanne.
23. Mandelbrot BB (1963) The variation of certain speculative prices. Journal of Business 36:394-419.
24. Lotka AJ (1926) The frequency distribution of scientific productivity. Journal of the Washington Academy of Sciences 12:317-323.
25. Zipf GK (1941) National Unity and Disunity. Principia Press, Bloomington.
26. Ijiri Y, Simon HA (1977) Skew Distributions and the Sizes of Business Firms. North-Holland, Amsterdam.
27. Osborne MFM (1959) Brownian motion in stock markets. Operations Research 7:145-173.

28. Canard N-F (1969) [1801] Principes d'Économie Politique. Edizioni Bizzarri, Rome.
29. Walras L (1874) Éléments d'Économie Politique Pure. F. Rouge, Lausanne (6th edition, 1926).
30. Poinsot L (1803) Éléments de Statique. Bachelier, Paris (8th edition, 1842).
31. Walker DA (2006) Walrasian Economics. Cambridge University Press, Cambridge.
32. Fisher, I (1930) The Theory of Interest. Augustus M. Kelly, New York.
33. Gibbs, JW (1902) Elementary Principles of Statistical Mechanics. Yale University Press, New Haven.
34. Samuelson, PA (1947) Foundations of Economic Analysis. Harvard University Press, Cambridge.
35. Föllmer, H (1974) Random economies with many interacting agents. Journal of Mathematical Economics 1:51-62.
36. Colander D, Holt RPF, Rosser, JB Jr (2004) The Changing Face of Economics: Conversations with Cutting Edge Economists. University of Michigan Press, Ann Arbor.
37. Kuhn TS (1961) The Structure of Scientific Revolutions. University of Chicago Press, Chicago.
38. Veblen T (1900) Preconceptions of economic science. Quarterly Journal of Economics 14:261.
39. Arrow KJ, Debreu G (1954) The existence of an equilibrium for a competitive economy. Econometrica 22:265-290.
40. McCauley JL (2004) Dynamics of Markets: Econophysics and Finance. Cambridge University Press, Cambridge.
41. Hodgson, GM (1993) Economics and Evolution: Bringing Life Back into Economics. University of Michigan Press, Ann Arbor.
42. Marshall A (1890) Principles of Economics. Macmillan, London (8th edition, 1920).
43. Veblen T (1898) Why is economics not an evolutionary science? Quarterly Journal of Economics 12:373-397.
44. Kauffman SA (1993) The Origins of Order: Self-Organization and Selection in Evolution. Oxford University Press, Oxford.
45. May, RM (1976) Simple mathematical models with very complicated dynamics. Nature 261:459-467.
46. Hartmann GC, Rössler OE (1998) Coupled flare attractorsa discrete prototype for economic modeling. Discrete Dynamics in Nature and Society 2:153-159.
47. Rosser, JB Jr (2001) Complex ecologic-economic dynamics and environmental policy. Ecological Economics 37:23-37.

Part IV

Comments and Discussions

Econophys-Kolkata II Workshop Summary

J. Barkley Rosser, Jr.

Department of Economics, James Madison University, Harrisonburg, VA, USA.
rosserjb@jmu.edu

Rather than a summary, these remarks should be viewed as an extension of ideas presented in my paper in this Workshop. In my paper I discussed the emergence of econophysics out of the past evolution of the relationship between economics and physics. In discussing the current status of econophysics in its relationship with economics, I introduced the distinctions between orthodoxy, mainstream, and heterodoxy in economics, with the first being a largely intellectually defined category, the second a largely sociologically defined category, and the third involving both intellectual and sociological elements in its opposition to the first two simultaneously. I then suggested that in its current state econophysics is more of the third category, heterodox, than of the first two in its status within economics. I also suggested that a fully successful move into the mainstream could lead to its ideas becoming part of a new orthodoxy, which could ironically lead to its disappearance as a distinct entity.

I am not wise enough to know whether this last outcome will happen or not, and whether it would be a good thing or not. If it were to happen, it would probably be a process that would take several decades, although some acceptance by at least parts of the non-orthodox mainstream is certainly going on at the current time. A more immediate question that faces econophysics, and practicing econophysicists, is the degree to which this still new discipline will succeed in establishing itself as a self-perpetuating and self-reproducing institutional entity in the intellectual universe. To the extent that it can do so, it may not matter that much how it is treated by the economics profession or relates to it. Of course, going back to the essentially sociological definition of econophysics given by Mantegna and Stanley, that would involve maintaining institutional connections with physics quite strongly, as the basic idea is that econphysics essentially involves physicists studying economics using ideas from physics. To maintain econophysics then as a distinct discipline in that sense must then involve them remaining physicists to some degree and not simply becoming economists, although some trend to the latter is probably inevitable in the longer run.

In looking at how the existing heterodox schools of economics persist, such as Marxism or Post Keynesianism or Austrianism, one can see that a complex set of interrelated institutional elements must be present. There must be universities or institutes where the followers of the school of thought must be located, where they are either openly supported and encouraged, or at least tolerated. In these locations they must be able to train students and support post-docs who can follow them and replace them and continue the existence of the school of thought. These in turn need to have definite outlets for their research in the form of friendly journals and societies, although getting research from the school of thought into broader outlets and broader acceptance is always a worthy goal.

I do not know enough about physics departments to say where econophysics is most firmly established, although I know that there are graduate students in such departments who are currently writing Ph.D. theses on topics that can only be described as economics or perhaps finance. There are also a number of research institutes where such students can find support as post-docs, with the Saha Institute of Nuclear Physics in Kolkata clearly being one of these, but not the only one. Such departments and institutes now seem to be operating in a number of countries, not just India and the United States, but at a minim in Italy, UK, Germany, France, China, and Japan as well, if the home countries of the participants at this workshop are any indication, although I am not in a position to judge how strongly established these outposts are in each of these respective countries or what their future prospects for funding for future research are.

One area where the emerging field clearly has strength is in establishing a firm presence in certain journals, especially physics journals. Leading ones include *Physica A*, currently edited by one of the neologizers of the term and leaders of the movement, H. Eugene Stanley, as well as *European Journal Physical Journal B*, *Physical Review E*, and some others to lesser degrees. Some influential papers in the field have also appeared in the more general and highly influential *Nature* and *Science*. These have certainly given a strong base of credibility for the movement and discipline.

Of course, a problem here is that in terms of influencing economists and the broader discipline of economics, few economists read these journals regularly. Indeed, unlike some multi-disciplinary journals that occasionally publish econophysics papers, such as *Quantitative Finance* and *Nonlinear Dynamics in Psychology and Life Science*, none of these papers are indexed in the *Journal of Economic Literature*, which is a main source for most economists of keeping track of relevant published research in their field. To the extent that the leaders of econophysics wish to influence economists and to have their research begin to be read and cited by more regular economists in their research, it would probably behoove the editors of some of these journals to work to get their journals so indexed. Some that publish especially large amounts of it, such as *Physica A*, should certainly do so, and should be accepted by the *Journal of Economic Literature* as being an ongoing and important outlet for economic research per se, even if most of it is being done by people who are nominally physicists.

Eventually it should probably be the goal of econophysicists to publish more directly in economics journals themselves. Some of this has occurred already in a subset of journals open to complex dynamics and multi-disciplinary approaches, such as *Journal of Economic Dynamics and Control* and *Journal of Economic Behavior and Organization.* Many of these papers have involved coauthors who are economists, and there does seem to be a trend towards this worthy and obvious collaboration. However, this observer suspects that one reason why physicists have held back from submitting their papers to economics journals is the somewhat longer time involved in getting back reports from referees and editors, along with a higher rate of rejections, that one encounters in the more specifically economics journals than in the more specifically physics oriented journals. It will probably be a long time before we see full blown econophysics papers in the core leading journals in economics, such as the *American Economic Review* or the *Journal of Political Economy,* if ever.

Regarding societies, I do not know of any that specifically exist. However, there are societies in economics that are clearly open to econphysicists, with the new Economic Science of Heterogeneous Interacting Agents (ESHIA) being one and the Society for Nonlinear Dynamics and Econometrics (SNDE) being another, as well as the more multi-disciplinary Society for Chaos Theory in Psychology and Life Sciences (SCTPLS) being another, this latter already having several econphysicists involved with it as does ESHIA. I would also note that it may be more likely that econphysicists will ultimately receive more welcome and influence in finance than in economics per se.

Let me conclude by reminding of the distinction between multidisciplinarity, interdisciplinarity and transdisciplinarity. The former involves researchers communicating while remaining clearly in their distinct disciplines. Interdisciplinarity involves carving out a narrow sub-specialty that takes small elements of each of the disciplines, such as a water economist. Econophysics could go this way, but it seems to me to be sort of a dead end. Rather I would recommend following the lead of the ecological economists who openly seek a deeper synthesis of the two disciplines involving a fuller knowledge and respect for both that can lead to more fruitful integration as implied by the transdisciplinary approach. It is this latter that I think is the way to go, and the emergnce of joint research and cooperation between practicing economists and physicists would seem to be the way to effecuate this outcome best.

Econophysics: Some Thoughts on Theoretical Perspectives

Matteo Marsili

The Abdus Salam International Centre for Theoretical Physics
Strada Costiera 11, 34014 Trieste Italy

Summary. I offer some personal considerations on the current trends and future perspectives in the field of econophysics.

The abundance of data recently made available in several disciplines has led to a new type of empirical science, pioneered by physicists, aiming at organizing such information into well defined statistical/empirical laws – or *stylized facts* (see e.g. [1, 2]). The interest in these laws is only partly related to their use in applications. It rather arises from the expectation that such empirical laws might be the manifestation of some fundamental or universal principle[1].

However, inferring fundamental laws – typically on the nature of the interactions – from empirical data (e.g. from correlations) is a subtle issue. While physics can draw on the knowledge of microscopic laws to make such an inference, in socio-economic sciences or finance things are much more complex. For example, as a corollary of market efficiency arguments, the presence of predictable patterns in financial time series (e.g. the well known January effect) signals the *absence* of traders exploiting such irregularities and their disappearance likely implies that traders have recognized such patterns and modified their behavior accordingly. In this case, empirical data on returns can at most say what agents are *not* doing. Physicists also know very well that in strongly interacting systems, correlations may extend much further than interactions do[2], so that a naive inference of the latter from the former might be misleading.

[1] It might be worth mentioning, in passing, that the difference between fundamental and applied science we are used to in Natural sciences acquires new dimensions in socio-economic sciences. For example, the laws of quantum mechanics would not change if they are used in a nano-device, but the "laws" of a financial market may change in response to those who try to exploit them.

[2] The typical example is the emergence of long-range order in spin models, where even sites far apart which do not interact share the same orientation of their spins.

While there are very good reasons why empirical studies are currently dominating the scene, on longer time-scales the contribution of theoretical modelling may become significant both for the quest of fundamental principles and in guiding empirical research (as it has been in theoretical physics). Theoretical models make the relation between interaction and measurable collective behavior explicit (even though not univocal, in general). The experience which statistical mechanics has accumulated in understanding how thermal fluctuations and interaction determine different states of matter is very valuable in this respect. As realized long ago [3], a similar statistical description of the states of interacting systems is possible even in other disciplines. Indeed, the relevance of such an approach – and of phase transitions as a consequence of interaction in particular – has also been realized by some economists[3] [4].

Indeed, while economists generally recognize the inadequacy of mainstream approaches and the need to properly account for agents' heterogeneity and stochastic effects [6], they still find the rough simplification of economic reality assumed by many models proposed by physicists unacceptable[4]. Indeed, in modelling socio-economic systems one faces the additional hurdle of dealing with a much more complex reality than that of physics. The indifference of statistical laws to microscopic details justifies only partly the ruthless simplification which models need to make of reality. In this respect, rather than insisting on realism, it might be better to propose and view such toy models as stylized realities where agents interact in well defined environments, capturing some key elements of the real systems. This might not leads to quantitative predictions for real systems, but it might provides a coherent picture which helps us think about these systems and sheds light on what new phenomena we can expect.

For example, both the Minority Game [8] and a phenomenological model of multi-asset markets [9] relate the emergence of non-trivial fluctuations in financial markets to the critical phenomena taking place in the proximity of a phase transition. In both cases, a simple argument suggests that markets spontaneously evolve towards the critical state. It roughly goes as follows: as long as the density of traders is small enough, the market offers reliable opportunities (of speculation or risk minimization) which may attract further investors. Newcomers, however, make these opportunities less reliable and drive the market closer to the critical point. Three points are worth noticing: *i)* both cases show the fallacy of the price taking assumption on which most of financial engineering is based; *ii)* the occurrence of an instability or critical point as the density of interactions increases is of the same nature of that discussed by May long ago [10] for complex systems in general and ecosystems

[3] However, the point has not been taken further than some generalizations of known results on the Ising model, and it has soon after been dismissed [5].

[4] Random exchange models proposed to explain the shape of the wealth distribution are an example [7].

in particular; and *iii)* this scenario is reminiscent of Self-Organized Criticality (SOC) [11]. As in SOC, the presence of two different processes acting at well separated time-scales is essential. However, there are two important conceptual differences: First the concept of avalanches does not play a significant role and second the tendency to converge to the critical point is not due to external driving but to endogenous forces: The agents themselves, responding to incentives, drive the system close to the critical point. Under such endogenous forces, the density of interactions increases.

Probably socio-economic phenomena are currently changing at a rate which is much faster than that at which we understand them. Many of the things which are traded nowadays in financial markets did not exist few decades ago, not to speak of internet communities. By contrast, Nature has been there since ever, but it has taken centuries to develop a reasonable understanding of little parts of it. In addition, we face a situation in which the density and range of interactions are steadily increasing, thus making theoretical concepts based on effective non-interacting theories inadequate.

The emphasis which empirical analysis has acquired recently is a natural response to the need of organizing our description of complex phenomena. It is not clear whether computational (e.g. agent based) approaches, which have gained so much momentum recently, are contributing in the same direction. Computational approaches have been very useful in physics because the knowledge of microscopic laws constrains theoretical modelling in extremely controlled ways. This is almost never possible for socio-economic systems. Rather than organizing a complex phenomenology in reduced set of aspects, computational approaches may expand the phenomenology in an uncontrolled way.

Phenomenological and toy models have been extremely useful in unveiling how interaction and thermal fluctuations compete in statistical physics. The example of the Minority Game suggests that such a theoretical approach may open a wider window on the behavior of complex systems making it possible to locate "point-wise" empirical findings in a broader theoretical landscape. Even though stylized, such a reference, well controlled theoretical framework helps us shape our way of thinking about strongly interacting socio-economic systems. Ultimately, it may provide sources of inspiration for guiding empirical research in mining huge data-sets in the quest for fundamental theories and organizing principles.

References

1. Mantegna RN, Stanley HE (2000) Introduction to Econophysics, Cambridge University Press
2. Albert R, Barabási A-L (2002) Rev. Mod. Phys. 74:47
3. Majorana E (1942) Scientia 36
4. Young HP, Durlauf SN (2001) (Eds.) Social Dynamics Brooking Institution

5. Durlauf SN (2005) Economic Journal, Royal Economic Society 115(504):F225-F243
6. Kirman AP (1992) J. Econ. Persp. 6:117-136
7. Chatterjee A, Yarlagadda S, Chakrabarti BK Eds. (2005) Econophysics of Wealth Distributions, Springer, Milan
8. Challet D, Marsili M, Zhang Y-C (2005) Minority Games, Oxford University Press, Oxford
9. Raffaelli G, Marsili M, e-print physics/0508159
10. May RM (1972) Nature 238:413
11. Bak P, Tang C, Wiesenfeld K (1987) Phys. Rev. Lett. 59:381

Comments on "Worrying Trends in Econophysics": Income Distribution Models

Peter Richmond[1], Bikas K. Chakrabarti[2], Arnab Chatterjee[2] and John Angle[3]

[1] School of Physics, University of Dublin, Trinity College, Dublin 2, Ireland. richmond@tcd.ie
[2] Theoretical Condensed Matter Physics Division and Centre for Applied Mathematics and Computational Science, Saha Institute of Nuclear Physics, 1/AF Bidhannagar, Kolkata 700064, India. bikask.chakrabarti@saha.ac.in, arnab.chatterjee@saha.ac.in
[3] Inequality Process Institute, P.O. Box 429 Cabin John, Maryland 20818-0429, USA. angle@inequalityprocess.org

In Econophysics-Kolkata I, we had concentrated on collecting the data for wealth and income distribution in various countries and the models (of econophysics) to comprehend or explain those distributions. As such, there have been several criticisms by economists on these modelling attempts (see, *Econophysics of Wealth Distributions*, Eds. A. Chatterjee, S. Yarlagadda, B. K. Chakrabarti, Springer-Verlag, Milan, 2005, and also *New Scientist*, March 12, 2005 issue, pp. 6-7). Recently, M. Gallegati, S. Keen, T. Lux and P. Ormerod (http://www.unifr.ch/econophysics; doc/0601001, Physica A [in press]) have published a criticism of several of these modelling efforts. In particular, the various wealth distribution model papers published in the Proceedings of Ecconophys-Kolkata I (mentioned above) have been referred to. In this Comments and Discussion section (of Econophys-Kolkata II) therefore we include a few responses to these criticisms, by physicists and social statistician.

1 Comments on models developed to describe and understand income and wealth distributions, by Peter Richmond

A number of different models have been proposed by members of the physics and economics communities to account for income distributions in society.

Agent models proposed recently from the physics community have attracted criticism from some economists on the grounds that the concept of exchange invoked at a micro level does not reflect the way the monetary system operates. This may be so, but from the physicists perspective it does seems that the exchange process captures essential elements that allow the calculation of distribution functions that offer scope for testing against empirical data. This is in the spirit of the early pioneers of studies of molecules that were captured as point particles. This also is clearly not true, but provided a basis on which to build. In response then the challenge is for the economists to develop models that perform better than the simple exchange models.

Expanding the philosophy to the models already proposed, it is certainly clear that some models do not yield a proper Pareto regime. So we can immediately say that they are inadequate and require further modification. As a result we are already seeing more generalized models that take account in more detail the nature of the exchange mechanism plus other factors such as weighted or preferential exchange.

This in turn will demand even more detailed comparison to empirical data. Such comparisons will need to be made not only with the stationary distribution function but also the time series and dynamics. Tagged data sets are now available in Japan and other countries may follow and release for research similar high quality datasets. Only when such data - that spans not only the rich but also the poor - is widely available will a rational debate over the merits of models be possible. Without such data we shall always be chasing shadows.

2 Comments on ideal-gas like models of income distribution, by Bikas K. Chakrabarti and Arnab Chatterjee

2.1 A simple game: pure gambling

Think of a (limiting) game like this: There are N players participating in a game, each having an initial capital of one unit of money. N is very large, and total money $M = N$ remains fixed over the game (so also the number of players N). The only move at any time is that two of these players are randomly chosen and they decide to share their total money randomly among them. As one can easily guess, the initial uniform (a delta function) distribution of money will soon disappear. Let us ask what the eventual steady state distribution of money after many such moves will be? At each move, a pair of randomly chosen players share a random fraction of their total money among themselves (nobody can create or destroy money — will go to jail if they do so!).

The answer is well established in physics for more than a century — soon, there will be a stable money distribution and it will be Gibbs distribution: $P(m) \sim \exp[-m/T]$; $T = M/N$ (A. A. Drăgulescu and V. M. Yakovenko, Eur. Phys. J. B 17 (2000) 723).

2.2 Modification I: Savings

Now think of a modified move in this game: each player 'saves' a fraction λ of the money available with him/her after each move and while going to the next move. Everybody saves the same fraction λ. What is the steady state distribution of money after a large number of such moves? It becomes Gamma-function like, while the distribution parameters of course depend on λ (A. Chakraborti and B. K. Chakrabarti, Eur. Phys. J. B 17 (2000) 167; M. Patriarca, A. Chakraborti, K. Kaski, Phys. Rev E. 70 (2004) 016104; see also J. Angle, Social Forces 65 (1986) 294; Physica A 367 (2006) 388; for a somewhat different model with similar results developed much earlier).

2.3 Modification II: Random savings

What happens to the eventual money distribution among these players if λ is not the same for all players but is different for different players? Let the distribution $\rho(\lambda)$ of saving propensity λ be such that $\rho(\lambda)$ is non-vanishing when $\lambda \to 1$. The actual distribution will depend on the saving propensity distribution $\rho(\lambda)$, but for all of them, the asymptotic form of the distribution will become Pareto-like: $P(m) \sim m^{-\alpha}$; $\alpha = 2$ for $m \to \infty$. This is valid for all such distributions (unless $\rho(\lambda) \propto (1-\lambda)^{\beta}$, when $P(m) \sim m^{-(2+\beta)}$). However, for variation of $\rho(\lambda)$ such that $\rho(\lambda) \to 0$ for $\lambda < \lambda_0$, one will get an initial Gamma function form for $P(m)$ for small and intermediate values of m, with parameters determined by λ_0 ($\neq 0$), and this distribution will eventually become Pareto-like for $m \to \infty$ (A. Chatterjee, B. K. Chakrabarti, S. S. Manna, Physica Scripta T106 (2003) 36; Physica A 355 (2004) 155; P. Repetowicz, S. Hutzler, P. Richmond, Physica A 356 (2005) 641).

2.4 What is the relation of these games to income distributions?

If we view the tradings as scattering processes, one can see the equivalence with these games. Of course, in any such 'money-exchange' trading process, one receives some profit or service from the other and this does not appear to be completely random, as assumed in the models. However, if we concentrate only on the 'cash' exchanged (even using the Bank cards!), every trading is a money conserving one (like the elastic scattering process in physics)!

But, are these tradings random? Surely not, when looked upon from individual's point of view: When I maximize my utility by money exchange for the n-th commodity, I may choose to go to m-th agent and for r-th commodity I will go to s-th agent. But since $m \neq n \neq r \neq s$ in general, when viewed from a global level, these trading/scattering events will all look random (although for individuals there is a defined choice or utility maximization).

Apart from the choice of the agents for any trade, the traded amount are considered to be random in such models. Some critics argue, this cannot be totally random as the amount is determined by the price of the commodity exchanged. Again, this can be defended very easily. If a little fluctuation over the 'just' price occurs in each trade due to the bargain capacities of the agents involved, one can easily demonstrate that after sufficient tradings (time, depending on the amount of fluctuation in each trade), the distribution will be determined by the stochasticity, as in the cases of directed random walks or other biased stochastic models in physics.

3 A Comment on Gallegati et al.'s "Worrying trends in econophysics", by John Angle

Mauro Gallegati, Steven Keen, Thomas Lux, and Paul Ormerod criticize a number of aspects of the literature on particle system models of personal income distribution in "Worrying Trends in Econophysics". Their fourth, last, and most emphatically made critique in "Worrying Trends" is of a particle system with zero-sum wealth transactions between particles and fixed aggregate particle wealth as a model of a distribution of personal income. In this class of model is the Inequality Process [1–23] (hereafter IP) derived from social theory beginning in the 1980's, and the more recent saved wealth model [24–31], a modification of the kinetic theory of gases. In the IP, a 0,1 discrete random variable determines which particle in a randomly selected pair wins. The losing particle gives up a fixed proportion of its wealth to the winner. In the saved wealth model, the stochastic driver of wealth exchange between particles is a continuous uniform random variate with support at [0,1]. There is clear asymmetry of gain and loss in the IP for any value of its parameter, a variable and obscured asymmetry in the saved wealth model, none when the saved wealth model subsumes the kinetic theory of gases [22].

This comment on Gallegati et al. seeks to persuade them that their critique of particle systems with zero-sum wealth transfers between particles is misguided, particularly in the case of the IP, which they do not explicitly discuss, although one of their number, Lux [32], has discussed it elsewhere. Gallegati et al. summarize their fourth critique in their abstract as:

> "Fourth, the theoretical models which are being used to explain empirical phenomena.
>
> The latter point is of particular concern. Essentially, the models are based upon models of statistical physics in which energy is conserved in exchange processes. There are examples in economics where the principle of conservation may be a reasonable approximation to reality, such as primitive hunter-gatherer societies. But in the industrialised capitalist economies, income is most definitely not conserved. The process of production and not exchange is responsible for this. Models which focus purely on exchange and not on production cannot by definition offer a realistic description of the generation of income in the capitalist, industrialised economies."

3.1 The issue is whether individual particle wealth converges before there is substantial change in aggregate particle wealth

Gallegati et al. note that aggregate personal wealth typically increases over time in industrialized market economies, dramatically so over centuries. They conclude from that fact that a particle system model with aggregate particle wealth fixed to a constant is irrelevant to economies that grow. However, the justification of a model that holds the sum of particle wealth constant is that proportional change in wealth at the particle level in a large population of particles is much faster than that in aggregate particle wealth. If so, particle wealth converges to an equilibrium before aggregate wealth changes. Angle [15] argues that distributions of wealth and income converge to their empirical equilibria well inside of a year even when violently perturbed, i.e., there is no reason to reject a particle system model because of the empirical irrelevance of a conservation constraint over decades or centuries. In the contemporary U.S., aggregate annual wage income remains at about the same fraction of gross domestic product year to year. Thus, the U.S. aggregate annual wage income varies at about the same rate as the gross domestic product, usually a few percentage points a year, a rate much slower than that of most individual annual wage incomes.

3.2 The IPs macro dynamics

The IP models wealth and income distribution dynamics better than Gallegati et al. think a model in its class can. The meta-theory from which the IP was abstracted asserts that the parameter of each IP particle represents that particles productivity in generating wealth. Changing that parameter toward greater productivity for all particles, changes the shape of the IPs stationary distribution toward that of the distribution of personal earnings of the most educated in an industrial labor force. Taking worker skill level and the unconditional mean of wage income as exogenous variables, the IPs macro model, the approximation to its stationary distribution, fits the distribution of wage income in the U.S. 1961-2001, conditioned on education (six levels) [22], even though the distribution changed substantially in those years.

3.3 "Realistic description" or parsimony?

But Gallegati et al. might well counter that taking the unconditional mean of wage income as exogenous, although accounting for change in the distribution of wage income, fails to explain the dynamics of unconditional mean wage income itself. The IP transfers wealth to particles, that by hypothesis and empirical analogue, are more productive of wealth, explaining upward drift in aggregate wealth without explicitly modeling that upward drift [22]. A drawback? No. If an IP-like empirical process becomes established, wealth flows to the more productive without central direction (cf. Adam Smith's invisible hand) or knowledge of how wealth is produced, supplying the more productive with the wherewithal to produce more wealth. The IP can work in a time homogeneous way throughout techno-cultural evolution. The idiographic model (cf. Gallegati et al.'s phrase "realistic description of the generation of income") that Gallegati et al. would prefer cannot. Such a model would be unparsimonious and generalize poorly.

3.4 The IP uses zero-sum competition to assess particle wealth productivity

The mechanism with which the IP transfers wealth to the more productive is the asymmetry of gain and loss. If the more productive lose less when they lose - because they rebound faster, or are treated more gently, or have more leverage - then an IP-like empirical process of competition will transfer wealth to the more productive, nourishing their more efficient production, and causing upward drift in aggregate wealth production. The IP implies faster upward drift in aggregate wealth production, the higher the level of productivity in a labor force, since its mechanism of wealth transfer to the more productive works more efficiently in a more productive labor force, i.e., an IP-like empirical process becomes an "attractor" over time [22].

Zero-sum competition is the perturbing mechanism that transfers wealth from the less productive to the more productive in the IP via the asymmetry of loss and gain of wealth. Zero- sum competition is how the IP learns about particle productivity. Zero-sum competition between particles is thus the IP's explanation of aggregate wealth production. If nature chose extreme parsimony for its algorithm to maximize wealth production, it may have chosen zero-sum competition as in the IP.

3.5 The IP as mainstream economics

Gallegati et al. write that they "... are economists who are critical of much of ... mainstream work in economics." The IP lets Gallegati et al. rejoin that mainstream while doing econophysics. The IP is as transaction-oriented as the most market-oriented, i.e., mainstream, economist could hope for in a model economy. The IP implies a number of mainstream propositions:

Mainstream Proposition	IP Implication
1. All distributions of labor income are right skewed with tapering right tails [33]; hence the impossibility of radical egalitarianism.	The IP generates right skewed distributions shaped like empirical distributions of labor income [22].

Mainstream Proposition	IP Implication
2. Differences of wealth and income arise easily, naturally, and inevitably via an ubiquitous stochastic process; cf. Gibrat's law of proportional effect [34] hence the impossibility of radical egalitarianism.	In the IP, differences of wealth arise easily, naturally, and inevitably, via an ubiquitous stochastic process [1,2,15,22].
3. Each workers earnings are closely tied to each worker's productivity; [part of economics since Aesop's fable of the ant and the grasshopper was the whole of economics]	In the gamma pdf macro model of the IP's stationary distribution, a particle's expected wealth is the product of the of the ratio of its productivity to the population harmonic mean of productivity multiplied by the unconditional mean of wealth [22].
4. The ratio of mean wages in two occupations remains constant, given no change in the productivity of labor in either occupation, despite fluctuation in the mean wage in the labor force as a whole.	In the gamma pdf macro model of the IP, the ratio of the expected wealth of two particles remains constant despite changes in the unconditional mean of wealth [22].
5. Labor incomes small and large benefit from a business expansion strong enough to increase mean labor income; "A rising tide lifts all boats"; there is a community of interest between rich and poor in prosperity; in radical egalitarianism these interests are opposed.	In the macro model of the IP, an increase in unconditional wealth increases all all percentiles of the stationary distribution of wealth by an equal factor [22].
6. Competition in the market creates wealth and transfers wealth to the more productive of wealth via transactions without central direction, in the 21st century as in the 18th when Adam Smith wrote in praise of the "invisible hand" of the market.	In the IP, competition between particles causes wealth to flow via transactions from particles that are by hypothesis and empirical analogue less productive of wealth to those that are more productive, nourishing wealth production, and explaining the upward drift in mean wealth without a) requiring knowledge of how wealth is produced or b) central direction, i.e, with extreme information efficiency [22]. This process can operate homogeneously over the entire course of techno-cultural evolution.

The Breadth of Empirical Phenomena the IP Presently Accounts For The IPs empirical explanandum includes:

- the sequence of shapes of the distribution of wage incomes of workers by level of education and why this sequence of shapes changes little over decades [22];

- the dynamics of the distribution of wage income conditioned on education as a function of the unconditional mean of wage income and the distribution of education in the labor force [16–18];
- why a gamma pdf is a useful model of the left tails and central masses of wage income distributions and why their far right tails are approximately Pareto pdfs [22];
- why the IPs parameters estimated from certain statistics of the wage incomes of individual workers in longitudinal data are ordered as predicted by the IPs meta-theory [15] and approximate estimates of the same parameters from fitting the gamma pdf model of the IPs stationary distribution to the distribution of wage income conditioned on education [22];
- the difference in shape between the distribution of income from labor and the distribution of income from tangible assets [9];
- the decrease of the Gini concentration ratio and the increasing dispersion of wealth and income over the course of technocultural evolution [1,2,22];
- the sequence of shapes of the distribution of personal wealth and income over the course of technocultural evolution [1,2];
- the universality of the transformation of hunter/gatherer society into the chiefdom, society of the god-king, with the appearance of storable food surpluses [1,2].

If one allows a coalition of particles to have a greater probability than 50% of winning, then the Inequality Process so modified reproduces features of the joint distribution of income to African-Americans and other Americans such as:

- the % minority effect on discrimination (the larger the minority, the more severe discrimination on a per capita basis) [7];
- the relationships among variables as specific as a) % of a U.S. states population that is non-white; b) median white male earnings in a U.S. state; c) the Gini concentration of white male earnings in a U.S. state; and d) the ratio of black male to white male median earnings in a U.S. state [7].

3.6 An invitation to collaborate

The IP implies the above six mainstream propositions of economics, has a wide scope of empirical explanatory power, and is radically parsimonious. I invite inspection of the IP and its explanandum. The data on which the IP has been validated are publicly available at low or no cost. The IP has only been validated with U.S. data; the Achilles heel of the IPs claim to universality? I invite replication on non-U.S. data. I invite Gallegati et al. to become IP collaborators by recognizing - perhaps via attempted disconfirmation - that the route to including economics among the sciences lies at the IPs frontier.

References

1. Angle J (1983) Proc. American Statistical Association, Social Statistics Section. pp. 395-400. Alexandria, VA: American Statistical Association
2. Angle J (1986) Social Forces 65:293-326
3. Angle J (1986) Proc. American Statistical Association, Social Statistics Section. pp. 259-263. Alexandria, VA: American Statistical Association
4. Angle J (1990) Proc. American Statistical Association, Social Statistics Section. pp. 279-284. Alexandria, VA: American Statistical Association
5. Angle J (1992) Journal of Mathematical Sociology 17:77-98
6. Angle J (1993) Journal of Mathematical Sociology 18:27-46
7. Angle J (1993) Proc. American Statistical Association, Social Statistics Section. pp. 197-202. Alexandria, VA: American Statistical Association
8. Angle J (1996) Journal of Mathematical Sociology. 21:325-358
9. Angle J (1997) Proc. American Statistical Association, Social Statistics Section. pp. 388-393. Alexandria, VA: American Statistical Association
10. Angle J (1998) Proc. American Statistical Association, Social Statistics Section. pp. 138-143. Alexandria, VA: American Statistical Association
11. Angle J (1999) Proc. 1999 Federal Forecasters' Conference. pp. 161-169. Washington, DC: U.S. Government Printing Office
12. Angle J (1999) Proc. American Statistical Association, Social Statistics Section. pp. 331-336. Alexandria, VA: American Statistical Association
13. Angle J (2000) Proc. American Statistical Association, Social Statistics Section. pp. 270-275. Alexandria, VA: American Statistical Association.
14. Angle J (2001) Proc. American Statistical Association, Social Statistics Section. [CD-ROM], Alexandria, VA: American Statistical Association
15. Angle J (2002) Journal of Mathematical Sociology. 26:217-270
16. Angle J (2002) Proc. 2002 Federal Forecasters' Conference. Washington, DC: U.S. Government Printing Office
17. Angle J (2002) Proc. American Statistical Association, Business and Economic Statistics Section. [CD-ROM], Alexandria, VA: American Statistical Association
18. Angle J (2003) Estadistica. 55:59-93
19. Angle J (2003) An entry in T. Liao, et al., (eds.), The Encyclopedia of Social Science Research Methods. Volume 2: 488-490. Thousand Oaks, CA: Sage
20. Angle J (2003) Proc. Conference of the Federal Committee on Statistical Methodology, November, 2003, Arlington, Virginia [http://www.fcsm.gov/events/papers2003.html].
21. Angle J (2005) Paper presented to the First General Scholarly Meeting, Society for Anthropological Science. February, 2005, Santa Fe, New Mexico
22. Angle J (2006) Physica A 367:388-414
23. Kleiber C, Kotz S (2003) Statistical Size Distributions in Economics and Actuarial Sciences. Wiley, Hoboken, New Jersey
24. Chakraborti A, Chakrabarti BK (2000) The European Physical Journal B 17:167-170
25. Chatterjee A, Chakrabarti BK, Manna SS (2003) Physica Scripta T 106:36-39
26. Chatterjee A, Chakrabarti BK, Manna SS (2004) Physica A 335:155-163
27. Patriarca M, Chakraborti A, Kaski K (2004) Physical Review E 70:016104
28. Chatterjee A, Chakrabarti BK, Stinchcombe RB (2005) Physical Review E 72:026126

29. Patriarca M, Chakraborti A, Kaski K, Germano G (2005) arXiv:physics/0504153v1
30. Chatterjee A, Chakrabarti BK, Stinchcombe RB (2005) in H. Takayasu (ed.), Practical Fruits in Econophysics, Conference Proceedings of the Third Nikkei Symposium on Econophysics, Tokyo, Japan, 2004. Tokyo: Springer
31. Drăgulescu AA, Yakovenko VM (2000) European Physical Journal B 17:723-729
32. Lux T (2005) in Chatterjee A, Yarlagadda S, Chakrabarti BK (eds.), Econophysics of Wealth Distributions, New Economic Windows Series, Springer Verlag, Milan, pp.51-60
33. Pareto V (1897) Cours dEconomie Politique. Vol. 2. Lausanne: Rouge
34. Gibrat R (1931) Les Inégalités Economiques. Paris: Sirey.